Springer
Tokyo
Berlin
Heidelberg
New York
Hong Kong
London
Milan
Paris

T. Ona (Ed.)

Improvement of Forest Resources for Recyclable Forest Products

With 118 Figures

Springer

Toshihiro Ona, Ph.D.
Associate Professor
Department of Forest and Forest Products Sciences
Graduate School of Bioresource and Bioenvironmental Sciences, Kyushu University
6-10-1 Hakozaki, Higashi-ku, Fukuoka 812-8581, Japan

ISBN 4-431-20250-1 Springer-Verlag Tokyo Berlin Heidelberg New York

Library of Congress Cataloging-in-Publication Data

Improvement of forest resources for recyclable forest products / T. Ona (ed.).
p. cm.
ISBN 4-431-20250-1 (alk. paper)
1. Forests and forestry--Congresses. 2. Forest management--Congresses. 3. Wood waste--Recycling--Congresses. I. Ona, T. (Toshihiro), 1963-

SD118.I44 2004
634.9--dc22

2004045231

Printed on acid-free paper

Springer-Verlag is a part of Springer Science+Business Media
springeronline.com

Typesetting: Camera-ready by the editor and authors
Printing and binding: Shinano Inc., Japan
SPIN: 10963845

PRINTED WITH SOY INK

Foreword

It is an honor and pleasure for me to write the foreword of this book comprising the proceedings of the Fourth Symposium on the Improvement of Forest Resources for Recyclable Forest Products. The symposium was organized by Dr. Toshihiro Ona, Associate Professor at the Graduate School of Bioresource and Bioenvironmental Sciences, Kyushu University, Japan, as part of the "Development of Forest Resources with High Performance for Paper Recycling" research project. This was supported by the Core Research for Evolutional Science and Technology (CREST) team at the Japan Science and Technology Agency Foundation (JST) and by Kyushu University. As a colleague of Dr. Ona, I commend his efforts in organizing the symposium and editing this book.

In the forest, there is a multitude of resources, including trees, herbal plants, fruits, fungi, mammals, birds, insects, fishes, reptiles, water, landscapes, and tourist attractions. Nowadays, even the environment is regarded as a kind of forest resource. These resources can provide a diversity of forest products, such as timber for buildings, pulp and paper, charcoal, herbal medicines, wild vegetables, animal protein, edible mushrooms, and nonwoody fibers. From these resources, major forest products are produced using various species of trees. For example, softwood is suitable as building material, while hardwood is suitable for furniture production; pulp and paper are produced from both softwood and hardwood. Therefore, forest locations and forest management methods should vary according to the tree species used for production of different forest products.

According to *Global Forest Resources Assessment 2000*, the total forest area in the world was estimated to be 3.9 billion ha, or 0.6 ha per capita, of which 95% was natural forest and 5% forest plantations. In 2000, the worldwide forest growing stock was estimated to be 386 billion m^3, and the aboveground woody biomass to be 422 billion tons. These figures indicate the importance of forests in terms of quantity of wood supply and absorption of carbon dioxide. Therefore, we must at least maintain, if not increase, the world's forest area and carbon dioxide storage in growing stock. At the same time, worldwide consumption of wood has gradually been increasing as wood consumption per capita increases with economic expansion and population growth. The Food and Agriculture Organization of the United Nations (FAO) predicted that wood consumption will increase by 15% in the next 10 years, mainly as a result of the anticipated economic development of China and population growth in developing countries in Asia and Africa. In such circumstances, we must develop science and technology that will make possible the sustainable utilization of forest products without reducing the forest area and growing storage.

The extension of product life and the expansion of product recycling play important roles in the sustainable utilization of forest products. With regard to forest conservation and the increase in carbon dioxide fixation, the repeated use of fiber in recycling is very important in the pulp and paper industry, while extending the lives of buildings and furniture are of great significance in the use of solid timber.

Within this book, the reader will find valuable topics of discussion aimed at finding and developing improved forest resources to better allow the recycling of pulp and paper fiber. As seen in the table of contents, the viewpoints taken in the discussions are broad and diverse, ranging from forest management to what is now referred to in Japan as ecomaterial science, which is concerned with the maximization of native characters and functions of materials simultaneously minimizing its disturbance impacts on the environment and the society considered from resources utilization to waste management. However, the contributors are consistent with regard to sustainable forest resource utilization and greater awareness of the environment. This book, which captures the content as well as the spirit of the symposium, will be valuable to all those who are concerned with the best use of the world's forest resources.

Kokki Sakai
Professor
Graduate School of Bioresource and Bioenvironmental Sciences
Kyushu University
August 2003

Table of Contents

Chapter 3 Anatomical Aspects

Chapter 5 Physical Aspects

Chapter 6 Biochemical Aspects

Chapter 1

Tree Selection

Overview of the Project "Development of Forest Resources with High Performance for Paper Recycling"

Toshihiro Ona
Graduate School of Bioresource and Bioenvironmental Sciences, Kyushu University, Hakozaki, Higashi-ku, Fukuoka 812-8581, Japan

Basic concepts

The concentration of atmospheric carbon dioxide is increasing and becoming a major problem. After the Kyoto protocol, afforestation is gaining support as one of the best solutions to solve this problem. Creating forest resources which are valuable for industry is an important factor to sustain afforestation and to maintain healthy forests. The recycling of forest resources will become possible only by realizing this.

Selecting superior trees is a vital factor in improving productivity if trees are grown for industry. Trees with superior growth can be easily selected using simple observation; selecting high quality trees (lumber quality) based on factors such as the potential yield and strength of the paper, requires much work and has been a problematic issue.

Many pulp and paper companies have tree plantation projects in many countries all over the world. Research into the selection problem led by Dr. Toshihiro Ona resulted in the development of a method to assess the quality of the whole trunk of a tree without felling. With this method, a rod-shaped sample, ie; Increment core sample, about 12 mm in diameter is cut from the trunk and analyzed using a Raman spectrometer. This method analyzes the chemical composition of the wood and predicts the quantities of wood constituents and their structures, pulp yield, strength and color (brightness), etc., in the pulp (paper). The analysis requires between five and ten minutes per sample.

Recycling waste paper (pulp fiber) is equally as important as recycling forest resources. Interest in environmental issues such as conserving energy and cutting carbon dioxide emissions has increased in recent years, as has demand for a higher proportion of waste paper (pulp fiber) to be collected and re-used. Many pulp and paper companies have responded to this demand in various ways including increasing the proportion of waste paper used. However, the paper manufacturing process degrades the pulp fiber, so that after repeated recycling, problems such as paper brightness and strength arise. These factors present an obstacle to increasing the proportion of waste paper used. Large quantities of agents are added to increase the strength of the paper and for other purposes during paper processing to solve these problems, but such agents also have their limits and adding chemicals is environmentally unsound. Under these circumstances, it is essential to drastically change the raw materials used and to produce (plant) and utilize forest resources more capable of withstanding recycling. Such actions would unite the recycling of forest with the recycling of waste paper and would establish the recycling of raw materials and a society with resource recycling systems.

To materialize this, several issues had to be solved.

One was the identification of forest resources (pulp fiber) which can withstand

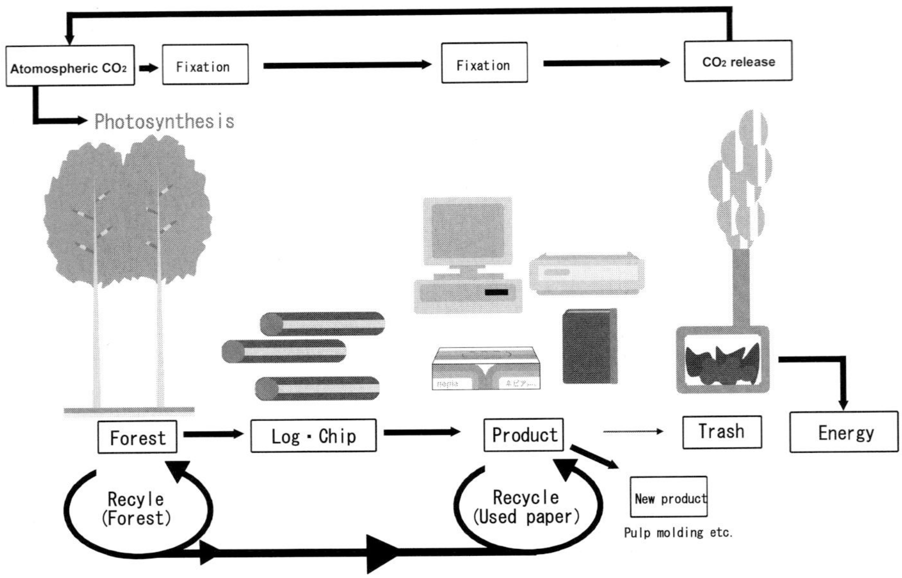

Figure 1 Research design.

recycling. The anatomical features and chemical compositions of fiber suitable for recycling have yet to be identified and require investigation. In particular, kraft pulp representing chemical pulp has an issue of strength reduction by repeated recycling, and the trees were searched and examined for no or less strength reduction by repeated recycling. On the other hand, high yield pulp, especially mechanical pulp is important for maximum utilization of resources. Although mechanical pulp has almost no problem of strength reduction by repeated recycling, it has an issue of low brightness by color reversion exposed to heat or light. For paper recycling, color reversion of paper consumes more bleaching chemicals to keep the paper brightness at a marketing level. Consequently, the trees were searched and examined for no or less color reversion by repeated recycling.

Another issue was the development of technologies to select superior trees with high performance for paper recycling. Identification and selection techniques of the required treeswere urgently required.

A third issue was cross-pollinating selected trees. Some hybrids grow much faster than the parent trees, which never happens with the same species. Producing hybrids of different species was another important target.

A fourth issue was to discover a mechanism of development of pulp fiber which has high performance for paper recycling inside the trees themselves. When trees grow from the immature stage to maturity, in general, wood fibers increase in length and then cease growth, resulting in a change in the fibers of the tree. Analysis of characteristic substances (anatomical features) in the immature and mature stages and revealing the causes of the development was required for future DNA recombination studies.

As explained above, the basic objective of this research was to establish a society with excellent recycling systems by fixing carbon dioxide in the air and conserving energy through pulp fiber both short term and long term (Figure 1).

Details of the research

The research could be separated into the following four segments (Figure 2).

1. Identification of pulp fiber
2. Development of selection techniques
3. Possibility of utilizing hybrids
4. Revealing the mechanism of development

(1) Identification of pulp fiber

We identified forest resources (pulp fiber) which have high performance for paper recycling. A few types of *Eucalyptus* and *Acacia* species and their hybrids were used as materials. Hand-made paper was produced by processing kraft pulp. The paper was pressed and dried, then the strength and other features of the paper was measured. These steps were repeated and the degradation in quality measured. The results were used to assess recycling performance. Additionally, differences between each raw material and positions inside the trunk were examined to determine the characteristics of pulp fiber with high performance for paper recycling. Also, differences between hybrid species in between different sub-species and the parent species were investigated to determine the possibility of improving quality through the use of the hybrids.

(2) Development of selection techniques

We established selection techniques for trees with high performance for paper recycling using a Raman spectroscopical method. The materials in item (1) were measured by a Raman spectrometer. Multivariate analysis of these results, the data of the anatomical features of wood fibers, and the chemical composition and strength of the paper after it had been recycled a specific number of times, gained in research item (1) were performed. Thus we established a quick and non-destructive measurement method using the Raman spectroscopy. We developed a device with an auto sampler to carry out analysis such as multivariate analysis automatically.

We used a quality prediction system which stipulates a pre-determined sampling position in order to take small samples from trees, rather than felling trees to predict the qualities of all species.

(3) Possibility of utilizing hybrids

We examined the possibility of utilizing cross-pollinated different species of trees (hybrids) and develop new types of trees with distinctive features.

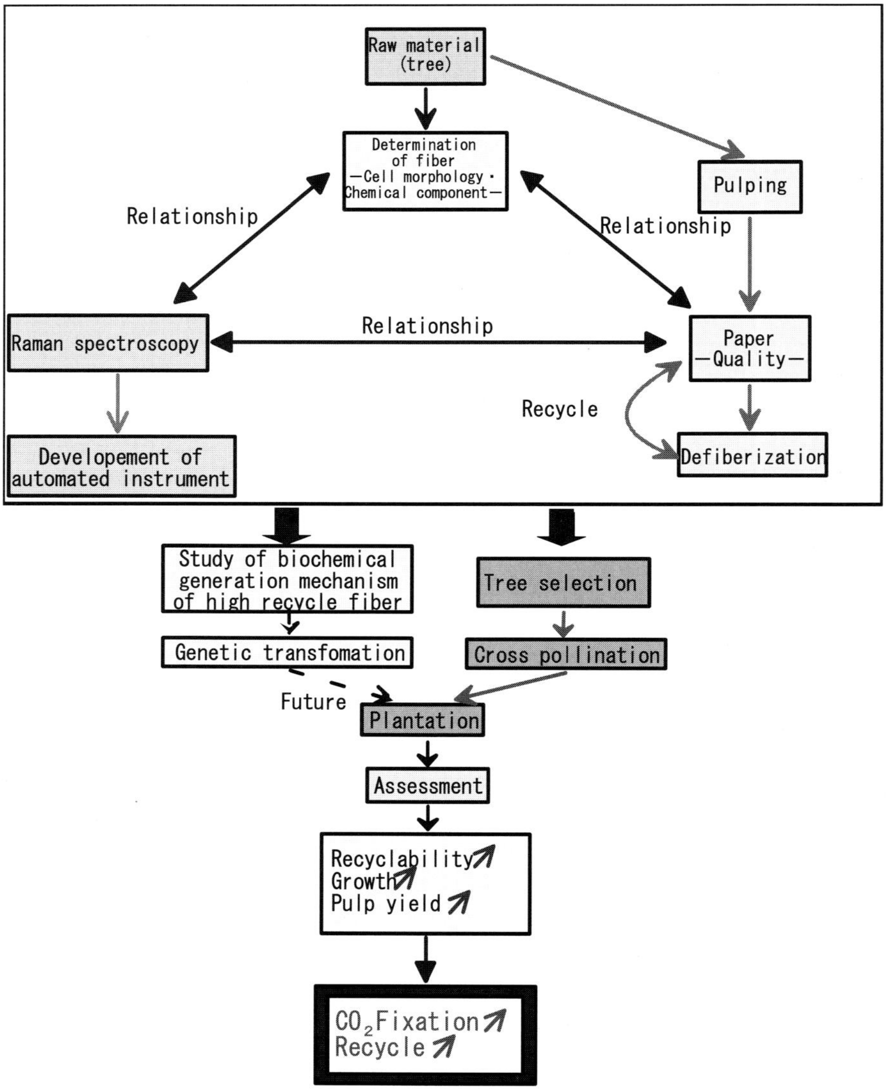

Figure 2 Research scheme.

(4) Revealing the mechanism of development

We examined the relationship between characteristic substances (chemical substances), growth stress (physical characteristics) and anatomical features in immature and mature stages. We examined the differences between mature trees and identical cloned trees rejuvenated by planting cuttings. In detail, regarding gene-related research, we researched DNA focusing on a potassium ion channel.

We analyzed phenol centering on substances. In research into growth stress, we investigated differences in stress inside trunks of living trees.

Near Infrared Spectroscopy: A Tool for the Rapid Estimation of Wood Properties

Laurence Schimleck[a], Robert Evans[b], David Jones[a], Richard Daniels[a], Gary Peter[c] and Christian Mora[d]

[a]*Warnell School of Forest Resources, University of Georgia, Athens GA 30605, U.S.A.*
[b]*CSIRO Forestry and Forest Products, Private Bag 10, Clayton South MDC, Victoria 3169, Australia*
[c]*School of Forest Resources and Conservation, University of Florida, Gainesville, FL 32611, U.S.A.*
[d]*Department of Forestry, North Carolina State University, 3125 Jordan Hall, Raleigh, NC 27695, U.S.A.*

Introduction

The wood properties of plantation grown trees are known to be quite variable. Owing to the variation that exists the potential for improving wood properties of plantation grown trees is great but to do so many trees have to be sampled. Typically methods employed to measure important wood properties are time consuming, expensive and often destructive, i.e. the tree has to be cut down, hindering the efforts of tree breeders to improve wood properties. To adequately measure the variation that exists rapid, cost-effective methods for measuring important wood properties of large numbers of samples are required.

Spectroscopic methods, including mid infrared (MIR) [1-2], near infrared (NIR) [3-8] and Raman [9-11], have been successfully applied to the rapid estimation of wood properties directly related to the chemistry of the wood including cellulose, lignin and pulp yield. Several studies have also demonstrated that NIR spectroscopy can be used to estimate a range of physical – mechanical properties of wood, including density and stiffness [12-17].

A great challenge in all spectroscopic studies is obtaining accurate calibration data and sufficient samples to describe the variation that exists. Recently we have used data provided by the SilviScan instruments [18-20] for calibration purposes. SilviScan-1 [18] and SilviScan-2 [19-20] have been specifically designed to test radial strips cut from increment cores and utilise a combination of scanning X-ray microdensitometry, X-ray diffractometry and image analysis to measure a range of wood properties including cell cross-sectional dimensions, air-dry density, microfibril angle (MFA) and stiffness (determined using SilviScan-2 diffractometric data and measured density). All properties are measured at high spatial resolution providing a profusion of calibration data at relatively low cost. In studies based on SilviScan data we have developed calibrations for the rapid estimation of wood properties of sections of radial wooden strips cut from *Pinus radiata* D. Don (radiata pine) increment cores [21-24]. In a more recent study we have used SilviScan data to developed wood property calibrations for *Pinus taeda* L. (loblolly pine) increment core samples obtained from a wide range of sites in Georgia, USA. We have also compared *P. taeda* wood property calibrations developed using NIR spectra obtained from green and air-dry radial strips [25]. The aim of this report is to summarise these studies and to provide a brief description of the measurement of wood properties using the SilviScan instruments and the collection of NIR spectra from the surface of radial wooden strips.

Materials and Methods

SilviScan analysis

Radial strips for analysis by SilviScan-1 and -2 were cut from increment cores or pith to bark radial samples using a twin-blade saw. Strip dimensions were 2 mm tangentially and 7 mm longitudinally, radial length was determined by the pith-bark length of the samples.

Air-dry density was measured in 50-micron steps using x-ray densitometry on SilviScan-1 [18]. MFA was measured in 1-mm steps on SilviScan-2 using scanning x-ray diffractometry [19-20]. Wood stiffness (at the same resolution as MFA) was determined using x-ray densitometry and x-ray diffraction data. All measurements were made in a conditioned atmosphere maintained at 40% RH and 20°C. Wood property averages were determined over 10-mm sections for correlation with the NIR spectra. A picture of SilviScan-2 is shown in Figure 1.

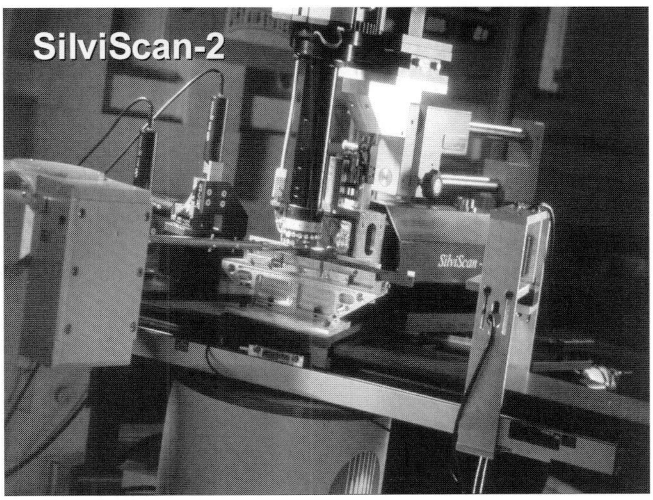

Figure 1 Picture of SilviScan 2.

Near infrared spectroscopy

NIR spectra consist mainly of overtone and combination bands of the fundamental stretching vibrations of O-H, N-H and C-H functional groups and contain chemical and physical information about a sample [26]. For the studies described in this report NIR diffuse reflectance spectra were obtained from the radial longitudinal face of each core sample using a NIRSystems Inc. Model 5000 scanning spectrophotometer. Samples were held in a custom made holder (Figure 2) [27]. A 5-mm x 10-mm mask was used to ensure collection of spectra from a constant area. The spectra were collected at 2 nm intervals over the wavelength range 1100-2500 nm. The instrument reference was a ceramic standard. Thirty-two scans were accumulated for each 10-mm section and the results averaged. One average spectrum was obtained per section. All NIR measurements were made in a conditioned atmosphere maintained at 40% RH and 20°C, matching the conditions under which the SilviScan measurements were made.

In studies based on green wood, NIR spectra were collected directly from the radial longitudinal and transverse face of radial samples cut from discs. It was observed that several samples were slightly twisted and a small gap between the spectrometer window and sample was occasionally observed permitting stray light to interfere with the NIR measurements. To minimize stray light the samples were tested in a light proof environment

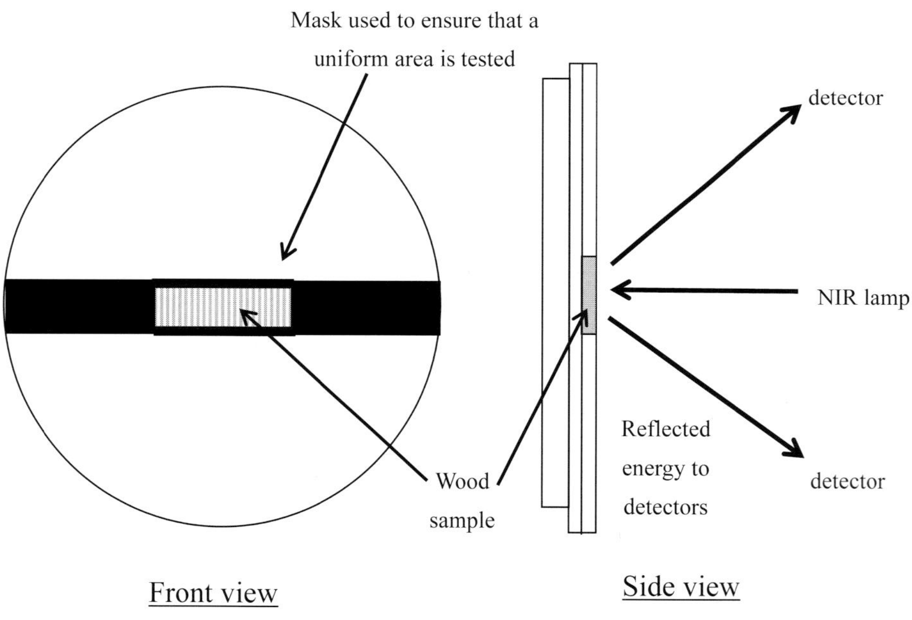

Figure 2 Sample cell used to hold radial wooded strips.

Calibration development

NIR spectroscopic analysis involves measuring the NIR spectra of a large number of samples, developing a regression calibration that links the spectra to the parameter of interest and then using the calibration and spectra of a new set of samples to validate the calibration [28-29]. In the studies that will be described Partial Least Squares (PLS) regression was used to develop the calibrations. The Standard Error of Cross Validation (SECV) (determined from the residuals of each cross validation phase), the Standard Error of Calibration (SEC) (determined from the residuals of the final calibration) and the co-efficient of determination (R^2) were used to assess calibration performance. The Standard Error of Prediction (SEP) was used to give a measure of how well a calibration predicts the parameter of interest for a set of unknown samples that are different from the calibration set.

Results and Discussion

P. radiata

In preliminary studies [21-24] that investigated the development of NIR calibrations using SilviScan data *P. radiata* samples were used. These samples are described in detail elsewhere [21]. The calibration set comprised 119 spectra (representing 8 radial strips) while the prediction set comprised 33 spectra (representing 2 radial strips). The samples were taken from breast-height in twenty-six-year-old trees, growing on a single site, that had been thinned at age fourteen. Five core samples were selected from trees that had been fertilised with nitrogen (200 kg/ha) and phosphorus (100 kg/ha) once after thinning [30]. The remaining five cores were chosen from trees that had been thinned but not fertilised. The cores were selected from trees that demonstrated different patterns of radial variation in MFA and represented the range of tree breast-height diameters. Calibrations were developed for several wood properties and were used to predict the wood properties of two strips (cores) which served as the separate test set. The results are summarized in Table 1 [21-24].

Table 1 Summary of calibrations developed for each P. radiata wood property. A total of 119 NIR spectra collected from the radial longitudinal face of 8 radial strips cut from increment cores were used for calibration development. The calibrations were tested on two radial strips, strip A (18 NIR spectra) and strip B (15 NIR spectra). Data were reproduced from Ref [21-24] with permission.

Wood	Calibration set			Strip A		Strip B	
Property	# factors	R^2	SEC	R_p^2	SEP	R_p^2	SEP
Air-dry density (kg/m^3)	5	0.90	23.5	0.95	34.8	0.92	62.3
MFA (deg.)	7	0.95	1.8	0.98	1.0	0.96	2.5
Stiffness (GPa)	8	0.97	0.9	0.99	1.6	0.98	1.2
Coarseness (μg/m)	9	0.91	31.0	0.80	51.5	0.91	76.6
Perimeter (μm)	6	0.68	4.6	0.41	6.1	0.57	5.0
Radial diameter (μm)	6	0.65	1.6	0.30	2.7	0.28	1.9
Tangential diameter (μm)	4	0.69	1.0	0.69	1.2	0.79	1.7
Wall thickness (μm)	7	0.89	0.2	0.88	0.2	0.91	0.5

The results presented in Table 1 demonstrate that it is possible to develop calibrations for air-dry density, MFA, stiffness and several tracheid morphological characteristics using NIR spectra obtained in 10-mm sections from the radial longitudinal face of radial wooden strips cut from *P. radiata* increment cores. The MFA calibration is shown in Figure 3.

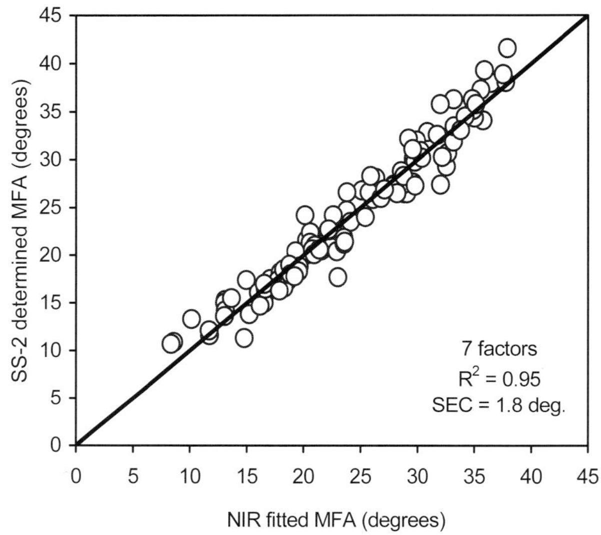

Figure 3 Relationship between measured microfibril angle (MFA) and NIR fitted MFA. The calibration was developed using 119 NIR spectra collected from the radial longitudinal face of 8 radial strips in 10mm increments. Note that the regression line has been plotted (reproduced from Ref [21] with permission).

The calibrations developed for air-dry density, MFA, stiffness, tracheid coarseness, tracheid wall thickness and tracheid tangential diameter all performed well when tested on strips A and B, while the radial diameter and perimeter calibrations did not.

The predicted air-dry densities of strips A and B are of particular interest. The SEP for strip B is much higher than that for strip A and over 2.5 times the size of the SEC. The range of air-dry densities represented by the calibration set was insufficient. Strip B had an average air-dry density that was 60.0 kg/m^3 higher than the average air-dry density of the calibration set. Because the lower density distribution in the air-dry density calibration set, the air-dry density calibration tended to underestimate the generally high air-dry density of strip B [23]. Fortunately the average air-dry density of each strip was known (determined from strip dimensions and weight – standard practice for SilviScan densitometry). The average air-dry density can be used to normalise the predictions of air-dry density of strips A and B. Radial variation of air-dry density, as determined by SilviScan-2 and predicted by the air-dry density calibration and then normalised, is shown for strip B in Figure 4. Note that normalization reduced the SEP from 62.3 to 28.5 kg/m^3.

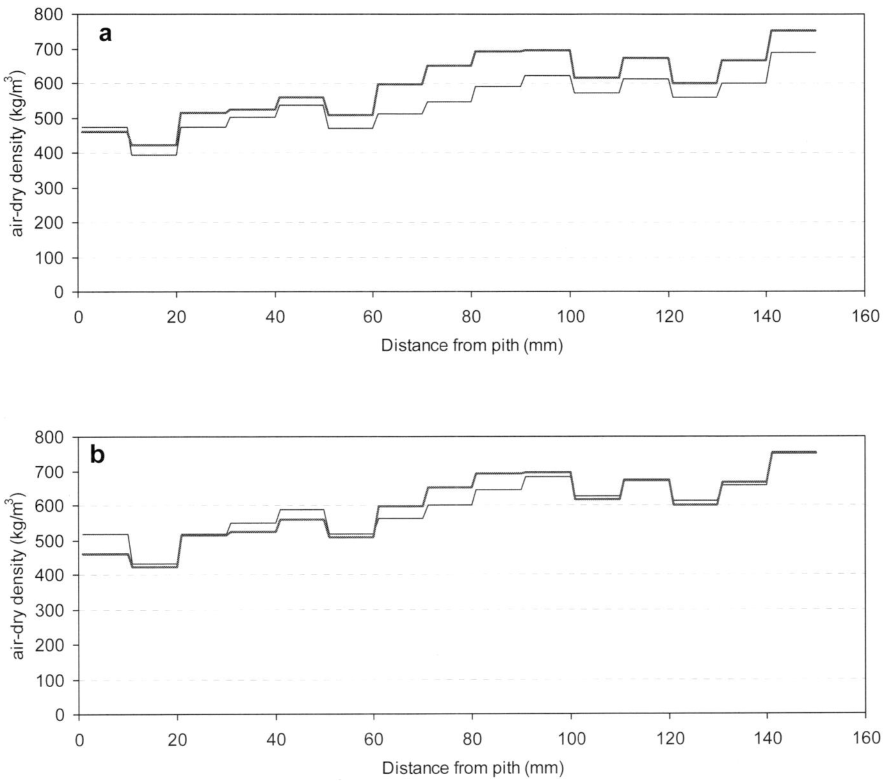

Figure 4 Plot of air-dry density pith to bark (radial) variation for P. radiata strip B before (a) and after (b) the NIR predictions of air-dry density were normalised. The thick line represents measured air-dry density. The thin line represents air-dry density predicted by the P. radiata air-dry density calibration developed using selected core samples and normalised using the known air-dry density of strip B (reproduced from Ref [23] with permission).

P. taeda

The *P. radiata* calibrations were based on a relatively small sample set (119 NIR spectra collected at 10 mm increments). If these calibrations were to be used operationally then it will be necessary to develop calibrations using several hundred samples that include as much genetic and site variation as possible. It is uncertain if the relationships would be as strong for large calibration sets.

Presently we are investigating what impact the use of large sample sets has on calibration statistics and the predictive performance of wood property calibrations. In this study 90 breast height (1.30 m) increment cores were collected from nine *P. taeda* plantations in Georgia, USA. For each geographic region where *P. taeda* is grown (Lower and Upper Atlantic Coastal Plain, and Piedmont) three plantations were sampled. The selected sites had a range of site qualities (low, medium and high). For each site ten trees with a range of breast height diameters were sampled.

Preliminary results for air-dry density, MFA and stiffness are presented in Figure 5. The calibrations developed for each wood property gave strong relationships with coefficients of determination (R^2) ranging from 0.85 to 0.95. The calibrations obtained show that it is possible to develop calibrations for important *P. taeda* wood properties using large data sets. The calibrations have slightly lower R^2 and slightly higher standard errors but this is to be expected considering the diverse origins of the calibration samples. The performance of these calibrations on a separate test set, comprising samples from sites not represented in the calibration, will be reported in the future.

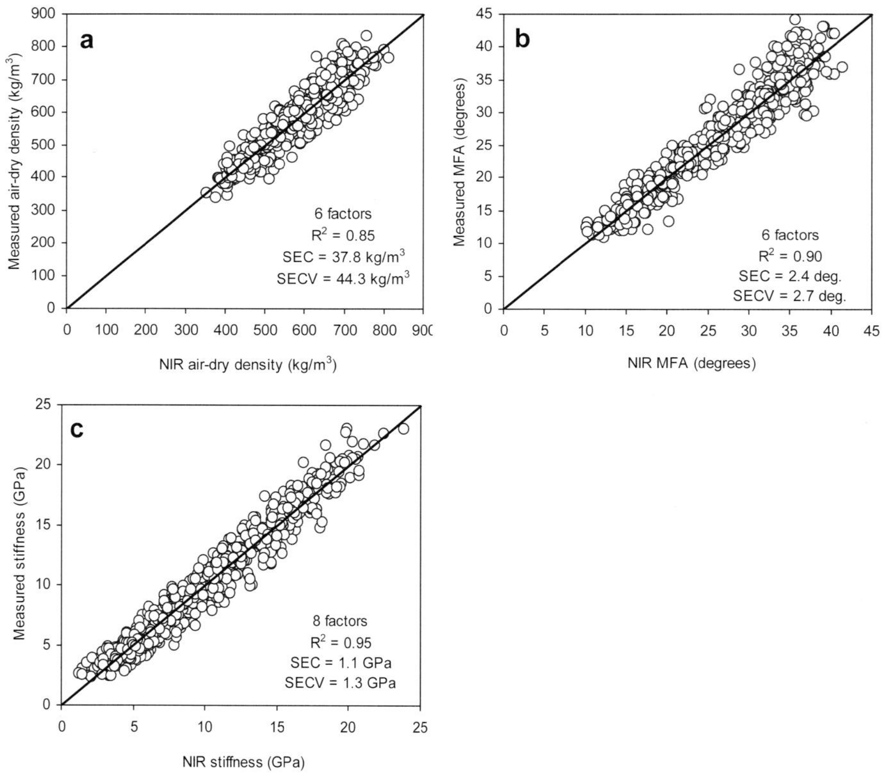

Figure 5 Relationships between measured values and NIR-estimated values for (a) air-dry density, (b) MFA and (c) stiffness. Calibrations were developed using 730 spectra representing 89 radial wood samples.

Application to green wood

The application of NIR spectroscopy to green *P. taeda* radial wooden strips has also been investigated. In this study *P. taeda* wood samples were obtained from a regeneration trial established by the North Carolina State Forest Nutrition Cooperative in 1979 on a poorly drained site in Williamsburg County, South Carolina, U.S.A. The study received a factorial combination of two levels of each site preparation, fertilization and herbicide treatment at establishment. A complete description of the treatments is given elsewhere [31]. Breast height (1.3 m) discs representing twenty trees from five different treatments were utilised in this study. All samples were frozen after sampling for storage. Radial samples representing pith to bark variation were cut from each breast height disc while frozen. The dimensions of the radial samples were 12.5 mm tangentially and 12.5 mm longitudinally, radial length was determined by the pith-bark length of each sample.

NIR spectra were collected from the surface of the 20 wooden samples in 10 mm increments when the samples were green and when air-dry. A sample holder similar to that shown in Figure 2 was used to collect the spectra. Radial strips were later cut from these samples for SilviScan analysis. Calibrations were developed for a range of wood properties using NIR spectra collected from 15 strips (three per silvicultural treatment) and then used to predict the wood properties of the remaining 5 strips (one per silvicultural treatment). The results for air-dry density, MFA and stiffness are summarized in Table 2 [25].

Table 2 Summary of air-dry density, MFA and stiffness calibrations developed from NIR spectra collected from the radial longitudinal face of P. taeda radial wooden strips when green and air-dry. Data were reproduced from Ref [25] with permission.

Wood	Calibration set				Prediction set	
Property	# factors	R^2	SECV	SEC	R_p^2	SEP
Green wood						
Air-dry density (kg/m^3)	3	0.85	39.2	35.2	0.74	57.6
MFA (deg.)	5	0.82	3.9	3.1	0.68	4.2
Stiffness (GPa)	5	0.88	2.3	1.8	0.81	3.0
Dry wood						
Air-dry density (kg/m^3)	4	0.90	36.0	29.6	0.90	32.8
MFA (deg.)	5	0.93	2.5	1.9	0.80	3.4
Stiffness (GPa)	6	0.95	1.6	1.1	0.92	1.9

The calibrations developed for each wood property using NIR spectra collected from green wood gave strong relationships with coefficients of determination (R^2) ranging from 0.82 to 0.88. The relationships were stronger for NIR spectra collected from dry wood (R^2 ranged from 0.80 to 0.93). While the green wood relationships were weaker they demonstrate that calibrations for air-dry density, MFA and stiffness can be developed using NIR spectra obtained in 10 mm sections from the radial longitudinal face of green *P. taeda* wood samples.

The results presented in this report demonstrate the following:

1. Data provided by SilviScan-1 and -2 can be used to develop calibrations for air-dry density, MFA, stiffness and several tracheid morphological characteristics. The calibrations developed for air-dry density, MFA, stiffness, tracheid coarseness, tracheid tangential diameter and tracheid wall thickness were found to provide good predictions of these properties when tested on NIR spectra obtained from two *P. radiata* increment cores.
2. Air-dry density, MFA and stiffness calibrations can be obtained using large calibration sets that include considerable genetic and site variation.
3. Air-dry density, MFA and stiffness calibrations can be obtained using NIR spectra collected from the radial longitudinal face of green wood.

In conclusion NIR spectroscopy has the potential to predict a range of wood properties in *P. radiata* and *P. taeda*. NIR spectroscopy offers a rapid, low cost alternative to traditional methods of analysis and could be used to identify trees for SilviScan analysis or wet chemical analysis. We estimate that 20-25 cores could be analysed per day (assuming a core radius of 100 mm). Automation of NIR analysis has the potential to double the number of cores analysed daily.

References

[1] Schultz, T. P. and Burns, D. A. Tappi J. 73 (1990) 209.
[2] Michell, A. J. Appita J. 47 (1994) 29.
[3] Birkett, M. D. and Gambino, M. J. T. Pap. S. Afr. November/December (1988) 34.
[4] Wright, J. A., Birkett, M. D. and Gambino, M. J. T. Tappi J. 73 (1990) 164.
[5] Garbutt, D. C. F., Donkin, M. J. and Meyer, J. H. Pap. S. Afr. April (1992) 45.
[6] Michell, A. J. Appita J. 48 (1995) 425.
[7] Schimleck, L. R., Raymond, C. A., Beadle, C. L., Downes, G. M., Kube, P. D. and French, J. Appita J. 53 (2000) 458.
[8] Raymond, C. A. and Schimleck, L. R. Can. J. For. Res. 32 (2002) 170.
[9] Ona, T., Sonoda, T., Ito, K., Shibata, M., Kato, T. and Ootake, Y. J. Wood Chem. Technol. 17 (1997) 399.
[10] Ona, T., Sonoda, T., Ito, K., Shibata, M., Kato, T. and Ootake, Y. J. Wood Chem. Technol. 18 (1998) 27.
[11] Ona, T., Sonoda, T., Ito, K., Shibata, M., Kato, T. and Ootake, Y. J. Wood Chem. Technol. 18 (1998) 43.
[12] Thygesen, L. G. J. Near-infrared Spectrosc. 2 (1994) 127.
[13] Hoffmeyer, P. and Pedersen, J. G. Holz Roh Werkst. 53 (1995) 165.
[14] Schimleck, L. R., Michell, A. J., Raymond, C. A. and Muneri, A. Can. J. For. Res. 29 (1999) 194.
[15] Meder, R. and Thumm, A. J. Near Infrared Spectrosc. 9 (2001) 117.
[16] Gindl, W. and Teischinger, A. J. Near Infrared Spectrosc. 9 (2001) 255.
[17] Meder, R., Thumm, A. and Marston, D. J. Near-infrared Spectrosc. 11 (2003) 137.
[18] Evans, R. Holzforschung 48 (1994) 168.
[19] Evans, R. Appita J. 52 (1999) 283.
[20] Evans, R., Stringer, S., and Kibblewhite, R. P. Appita J. 53 (2000) 450.
[21] Schimleck, L. R. and Evans, R. IAWA J. 23(2002) 225.
[22] Schimleck, L. R. and Evans, R. IAWA J. 23(2002) 217.
[23] Schimleck, L. R. and Evans, R. Appita J. 56(2003) 312.
[24] Schimleck, L. R. and Evans, R. Holzforschung (accepted) (2003).
[25] Schimleck, L. R., Mora, C. and Daniels, R. F. Can. J. For. Res. (accepted) (2003).

[26] Kaye, W. Spectrochim. Acta. 6 (1954) 257.
[27] Schimleck, L. R., Evans, R. and Ilic, J. Can. J. For. Res. 31 (2001) 1671.
[28] Martens, H. and Næs T. Trends Anal. Chem. 3 (1984) 204.
[29] Thomas, E. V. Anal. Chem. 66 (1994) 795A.
[30] Nyakuengama, J. G., Downes G. M. and Ng, J. IAWA J. 23 (2002) 431.
[31] Nilsson, U. and Allen, H.L. For. Ecol. Manage. 175 (2003) 367.

Forest Diversity and Pulp Quality: Some Tools for Wooden Raw Material Strategies for the Pulp and Paper Industry

Denilson da Silva Perez[a], Guillaume Chantre[b] and André Thémelin[a]
[a]*Wood-Process Laboratory, Domaine de l'Etançon, AFOCEL, 77370 - Nangis, France*
[b]*South West Unit, Domaine de Sivaillan, AFOCEL, 33480 - Moulis en Médoc, France*

Introduction

In order to increase pulp quality, to reduce mill operation fluctuations, and to decrease wood raw material costs, a clear tendency for the pulp and paper industry to operate single wood species coming from fast growing planted forests is nowadays observed [1]. Selected elite trees based on important breeding programs are used to compose the forests in order to reduce rotation time and to increase both pulp yield and quality. Eucalypt in South America and Australia, radiata pine in North America, and maritime pine in Southern Europe are some examples of this approach.

However, in several countries, pulp mills are still required to operate with mixture of hardwoods or softwoods coming from native forests. This is the case in France, the only exception being the mills in South-West which wood supply is exclusively composed of maritime pine. The diversity of wood species requires a complete characterization of the native forest potential in terms of pulp and paper quality. Moreover, when a pulp mill operates with mixtures, a fundamental question is permanently to be solved: how to optimize the mix of wood chips as a function of the wooden resources available within the mill's supply area and the end-product quality?

AFOCEL's major goal is to improve the wood supply competitiveness of French pulp and paper industry. In this paper, we report the work we have developed concerning the characterization of French wood species, the behavior of wood species mixtures and the maritime pine breeding program.

Results and Discussion

Singles wood species pulp and paper potential

A complete characterization of softwoods and hardwoods used by pulp mills in France was carried out concerning the wood logs characteristics (origin, moisture content, bark content, diameter distribution), wood chemical composition, kraft cooking behavior, fibers morphology and some strength paper properties. Data were organized to create an "ID card" for each wood species, allowing the mills to better organize their wood supply based on wood, pulp, and paper technical aspects.

Wood species are not the only source of heterogeneity of raw material for the pulp and paper industry. Considerable differences are due to the tissues (heartwood or sapwood, earlywood or latewood, normal or tension wood) and cell (fibers, tracheids, parenchymas or vessels) forming the wood [2]. Two approaches have been used to improve the knowledge on this topic: *i.*) the study of the characteristics and performances of different tissues or fibers; *ii.*) the distribution of these structures inside the trees. An example of clear differences between fibers characteristics of earlywood-latewood and heartwood-sapwood for maritime pine is given in Figure 1 [2, 3]. These differences surely affect the pulp quality during pulping and papermaking. The modelling of tissues distribution along the trees allows us to predict the fibers morphology and the pulp strength properties as function of trees characteristics such as diameter of trees and logs and the age (Figure 2) [4].

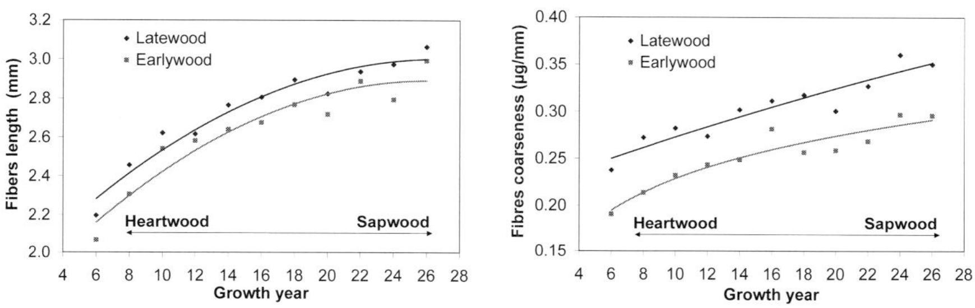

Figure 1 Fibers length of earlywood-sapwood and heartwood- sapwood for maritime pine.

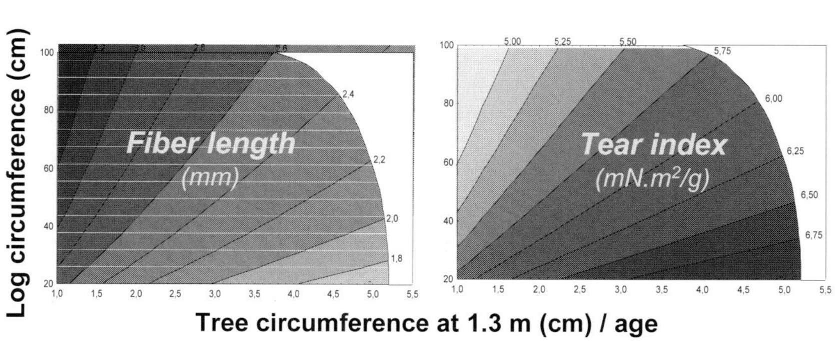

Figure 2 Fiber morphology (left) and tear index (right) predictions from maritime pine trees data [4].

Kraft cooking of wood species in mixture

The behavior of mixtures hardwood or softwood species was studied by using the Scheffé's polynomials [5, 6]. The property to be modeled is expressed as a function of component proportions:

$$y=\sum_{i=1}^{q} a_{ia}x_i+\sum\sum_{i\le j}^{q} b_{ij}x_ix_j+\sum\sum\sum_{i\le j\le k}^{q} c_{ijk}x_ix_jx_k+etc... \quad (1)$$

Different possibilities are then possible for the mathematical models: *linear* (no interactions), *quadratic* (second orders interactions, x_ix_j), or *cubic* (third order interactions, $x_ix_jx_k$, $i \neq j \neq k$).

Both hardwoods and softwoods mixtures were studied [7]. Hardwoods mixtures were based on poplar and beech, oak (mix 1) or chestnut (mix 2) being added as the third component. Douglas fir and Norway spruce composed the basis for the softwoods mixtures and the third component was either maritime (mix 3) or Scott (mix 4) pines. Only the most important results for hardwoods are shown here. The complete study is described in details in reference 7.

As it can be seen in Figure 3, the pulp yield of mixture poplar-beech-oak can be described by a quadratic model, showing strong two-by-two interactions between the components. The addition of chestnut instead of oak leads however to a linear model [6]. The introduction of chestnut or oak in a beech-poplar mix leads to intriguing interactions on the coarseness model. The third-order model lead to an interesting system where the higher coarseness is obtained with a mix which composition is 26% of oak, 38% of beech, and 36% of poplar. Finally, tear index is considerably improved when oak is used as the third component of the mixture.

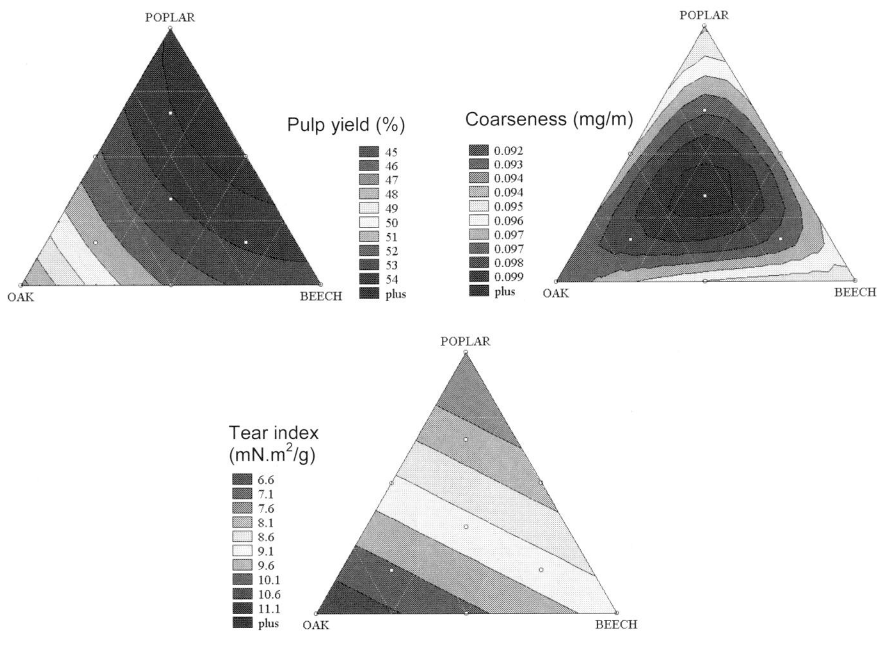

Figure 3 Pulp yield (upper left), coarseness (upper right) and tear index (lower) behaviour of poplar-beech-oak mixture.

Trees selection and breeding programs

The maritime pine breeding program started in the 60's with the selection of 360 elite trees (G0) from different origins, in order to guarantee a broad genetic basis for the characters of selection such as the adaptation to the Landes forest conditions, the growth rate and mainly the tree straightness. At the beginning of the 70's, the first generation of improved varieties (G1) was created by crossing pollens from these selected trees. Since the beginning of the program, at least 100,000 ha were planted with seeds from this first generation of improved maritime pine (10% of Landes forest). Important gains were obtained on volume (+15%) and on the trees straightness (+20%). These improvements had a direct impact on pulp yield, due to an increase in cellulose content associated with the reduction in lignin content estimated by NIRS (Near Infrared spectroscopy) [8].

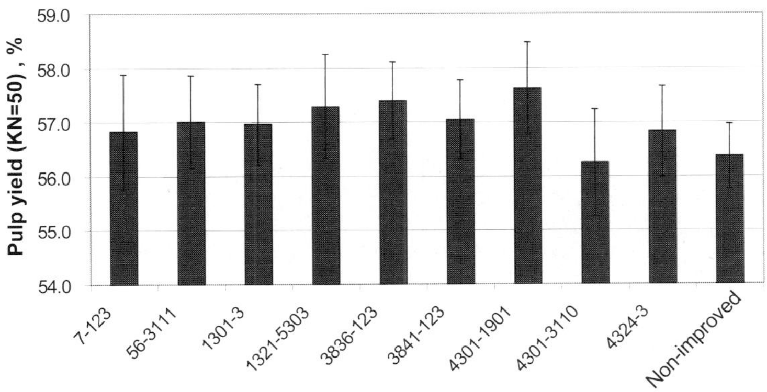

Figure 4 Pulp yield for G1 improved families (father-mother) of maritime pine.

The second generation of improved maritime pine (G2) was created from 40 elite G0 trees and 180 G1 trees in 1988. The estimations of gains with G2 trees are +30% on both volume and straightness in comparison to the G0 generation. A reduction of 10 years (from 46 to 36) in the rotation time would be possible since the production will increase from 10 m³/ha/year (with direct semis) to 13 m³/ha/year (with seeds). Since 1999, second generation of improved maritime pines are used in new plantations in Landes forest.

The selection of elite trees for the third generation of improved maritime pine will be based on modern genetic tools, such molecular markers and QTL (quantitative trait loci) identification. Different European projects (GENIALITY, UDH MAP, GEMINI) have been carried out to achieve a complete genetic characterization of maritime pine and to establish relationships with pulp traits. NIRS is a fundamental tool in this screening work, since more than 6,000 trees from different origins, clonal and factorial tests have been tested. One of the results from this effort was the preliminary identification of QTL connected with growth rate, chemical composition, physical properties and pulp yield [9]. As an example, some QTLs related to cellulose and lignin contents and pulp yield located in the linkage group 7 are shown in Figure 5.

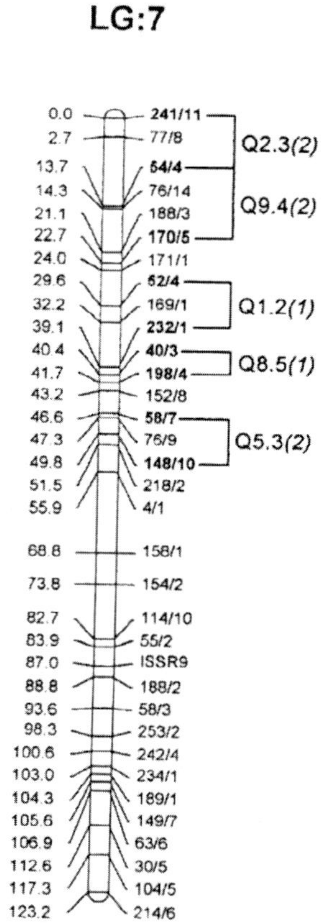

Figure 5 QTLs connected with cellulose content (Q1), lignin content (Q2), color (Q5), density (Q8) and diameter (Q9) located on LG7 of maritime pine [8].

References

[1] Rooks, A. Solutions 11 (2002) 29.

[2] Da Silva Perez, D. and Fauchon, T.: Wood quality for pulp and paper. In: Wood quality and its biological basis, Ed.: Arnett, J. R. and Jeronimidis, G. chap. 7 (2003) 157-186.

[3] Chantre, G. and Da Silva Perez, D. Natural variability of maritime pine fibres and industrial consequences (in French). VI Colloquium ARBORA "From planted forests to tomorrow's industry : Maritime pine from fibers to materials", proceedings, mars (2002) 36-61.

[4] Chantre, G., Bouvet, A., Sens, D., Robin, E. and Bongrand, O. Modeling kraft fiber morphology and paper properties from forest data: the example of maritime pine thinning logs, 2000 TAPPI Pulping / Process & Product Quality Conference Proceedings, Boston (2000).

[5] Scheffé, H. J. Royal Stat. Soc. (1958) 344.
[6] Scheffé, H. J. Royal Stat. Soc. (1963) 235.
[7] Da Silva Perez, D., Medina, S., Fauchon, T., Lecourt, M. and Deleuze-Brezins, C. Kraft cooking behaviour of wood species in mixture, 4th Biannual Johan Gullischen Colloquim, Espoo (Finland), proceedings, September (2003) 93-102.
[8] Da Silva Perez, D., Guillemain, A., Bouvet, A., Rodrigues, J. C. and Chantre, G. Near Infrared spectroscopy as a tool for the study of genetic determinism in maritime pine (*Pinus pinaster* Ait), 12th Intl. Symp. Wood Pulp Chem., Madison (USA) June (2003).
[9] Markussen, T., Fladung, M., Achere, V., Favre, J.-M., Favre-Rampant, P., Aragones, A., Da Silva Perez, D., Espinel, S. and Ritter, E. Silvae Genetica 52 (2003) 8-15.

Feasibility Study of Tree Selection for High Pulp Yield, Brightness and Recyclable Chemithermomechanical Paper Production Using *Eucalyptus globulus*

Toshihiro Ona[a], Mari Tateishi[a], Hideo Nozaki[b], Teruyuki Seino[c], Seung-Lak Yoon[d], Shinichi Isaji[e] and Yasuo Kojima[e]

[a]*Graduate School of Bioresource and Bioenvironmental Sciences, Kyushu University, Hakozaki, Higashi-ku, Fukuoka 812-8581, Japan*

[b]*Faculty of Agriculture, Kyushu University, Hakozaki, Higashi-ku, Fukuoka 812-8581, Japan*

[c]*NEDO Fellow, Institute for Environmental Management Technology, AIST Tsukuba West, Tsukuba 305-8569, Japan*

[d]*Department of Science and Engineering, Chinju National University, Chinju 660-758, Korea*

[e]*Graduate School of Agriculture, Hokkaido University, N9-W9, Kita-ku, Sapporo 060-8589, Japan*

Introduction

Eucalyptus is one of the most widely utilized species for tree plantation, especially intended in the use for chemical pulp production [1]. *E. globulus* is a fast growing species in a moderate zone with an 8-10 year-rotation period [2]. The tree breeding programs are aiming to improve growth, pulp yield and strength in pulp and paper products because the elite tree selection contributes to reduce energy consumption, CO_2 emission and pulp and paper cost.

The construction of an effective carbon loop, higher pulp yield and paper recycling rate are needed. Mechanical pulp has higher pulp yield than kraft pulp, which is suitable for this objective. Furthermore, the one of the mechanical pulps, chemithermomechanical pulp (CTMP), has an advantage that the paper strength does not decrease rapidly during recycling. However, the CTMP papers generally have a disadvantage of color reversion by heat exposure [3]. The degree of color reversion is crucial for paper recycling because it determines requirement amount of bleaching agents relating to pulp fiber damage during the bleaching process.

In this paper, we performed the feasibility study of elite tree selection of *E. globulus* in respect to high growth rate, pulp yield, brightness and low color reversion by heat exposure for recyclable CTMP paper production using seven individuals from Western Australia plantation site.

Experimental

Materials

Total 7 individuals of *Eucalyptus globulus* were similarly silvicultured and grown in two plantation sites of Manjimup (MJP) and Albany (ABN) in Western Australia (age 9.5).

CTMP

The wood chips were soaked in 5% sodium sulfite solution of pH 9.9 at 70°C for 1 h following at 20°C for 11 h. The primary refining was performed at 135°C for 3 min with preheating for 5 min by a defibrator (Metso Defibrator Type D, Helsinki, Finland). The second refining was by a PFI mill (STFI, Stockholm, Sweden) under pulp consistency of 20% at clearance of 0.5 mm. The CTMP was screened by a laboratory flat screen with an 8-cut plate to make hand sheets. The CTMP hand sheets were exposed to dry heat at 105°C for 24 h. The ISO brightness of sheets before and after dry heat exposure was measured by a spectro whiteness color meter (Suga Test Instruments SC-10W, Tokyo, Japan). Pulp productivity was calculated by screened pulp yield x basic density.

Wood property determination

The contents of holocellulose, α-cellulose, hemicellulose, lignin, and extractives and alkali-extractives were determined by using a small-scale method [4]. Basic density was determined by a water immersion method [5] followed by freeze-fracture and -drying, and expressed as dry weight/green volume [kg m^{-3}] [6]. Molar composition of neutral sugars of hemicellulose (glucose, xylose, galactose, arabinose and mannose) was quantified by hydrolysis of holocellulose by trifluoroacetic acid and high performance liquid chromatography [7].

The characterization of lignin structure including syringyl/guaiacyl (S/G) ratio was determined by pyrolysis according to a previous report [8] described in a paper by Tateishi et al. in this book.

Fiber morphology (length, diameter, lumen diameter and wall thickness) was assessed by the calibrations created by Raman spectroscopy modifying previous reports as follows [9, 10]. Raman spectra were collected using a Nicolet Raman 960 spectrometer (Nicolet Instrument, Madison, WI, USA) equipped with an InGaAs detector, and an Nd: YAG laser operating at 1064 nm. Tree samples (20 mesh - 0.840 mm opening - pass) were packed into nuclear magnetic resonance tubes and spectra were collected using 180°backscattering at 500 mW, 4 cm^{-1} resolution and 256 scans under 22°C using a thermal box [11]. Spectra of 100-1800 cm^{-1} Raman shift were used. The 1st derivative transformation and the multiplicative scatter correction of Raman spectroscopic data, and the partial least squares (PLS) regression were performed by using the software of Unscrambler 7.5 (Camo AS, Trondheim, Norway). The number of factors used in the PLS calibration model was decided by full-cross-validation method and useful frequencies were selected by Jack-knifing method [12]. A total of 226 samples for calibration sets were utilized from various parts in the trunk of two individual each of *E. globulus* and *E. camaldulensis* used in the previous studies [9, 10].

Statistical analysis

Correlation coefficients between CTMP and wood properties were obtained by the analysis of variance using a statistical computer software, SPSS 11.0.1J (SPSS Inc., IL, USA).

Results and Discussion

Tree selection by pulp property (conventional method)

The tree propeties of *E. globulus* and elite tree candidates were shown in Table 1.

Table 1 Tree data.

Property	Volume (m^3)	Pulp productivity ($kg \cdot m^{-3}$)	Virgin brightness (%ISO)	Heated brightness (%ISO)
MJP1	0.258	302	65.2	63.5
MJP4	0.312	372	65.9	64.0
ABN38	0.242	396	67.9	64.9
Min	0.231	302	60.3	57.8
Ave	0.262	358	65.0	62.5
Max	0.312	405	68.5	65.1

Large between-tree variation was observed in tree volume under the bark and pulp yield although they were silvicultured and grown equivalently. As expected, the CTMP yield was higher than kraft pulp [13]. The *E. globulus* CTMP owned high brightness without bleaching using chemical agents. Surprisingly, the highest brightness level in utilized *E. globulus* CTMP is almost equal to a bleached commercial copy paper manufactured from kraft pulp. In brightness, large between-tree variation was observed similar to volume and pulp yield. This implies that *E. globulus* CTMP from selected trees can be used in various types of papers without a bleaching process and minimize bleaching facility in a pulp production mill.

Consequently, the elite tree selection is encouraged for the effect utilization of forest resources, minimizing energy and chemical agent consumption and CO_2 emission.

The change of ISO brightness by dry heat exposure was examined and the results were shown in Table I. The color reversion of *E. globulus* CTMP by dry heat exposure was very weak in each individual and the CTMP brightness is very stable against dry heat exposure. This suggests that the papers manufactured from *E. globulus* CTMP allows the paper storage under high temperature without light and humidity. Furthermore, virgin brightness positively correlated to heated brightness with correlation coefficient of 0.978, which was significant at more than 1% level although no relationship between screened pulp yield and brightness was observed. The selection of individuals producing brighter virgin pulp will permit the paper more recyclable since severe pulp fiber damage by bleaching can be avoided by the reduction of bleaching agent amount.

An individual of MJP4 can be selected as an elite tree providing 1.19 times higher volume (growth), 1.04 higher pulp yield, 1.01 times higher virgin brightness and 1.02 times higher heated brightness compared with the average of utilized 7 individuals. The tree selection will help the construction of efficient carbon loop.

Tree selection by wood property (increment core method)

Tree selection is preferably performed by using an increment core, which retains the tree for future breeding, pollination and propagation. To achieve this, significantly high relationship between pulp and wood properties is required. The relationships for pulp yield and for brightness were shown in Table 2.

Table 2 Correlation coefficient between pulp and wood properties.

Pulp property	Wood property	Correlation coefficient
Screened pulp yield	α-Cellulose	0.905 ***
	Hemicellulose	− 0.718 **
	Holocellulose	0.685 **
	Xylose	− 0.684 **
	trans-Sinapaldehyde	0.681 **
	Syringol	− 0.651 *
	Homosyringaldehyde	0.664 *
	Glucose	0.643 *
	trans-Propenylsyringol	0.640 *
	Mannose	0.638 *
	trans-Sinapyl alcohol	− 0.621 *
	Acetoguaiacone	0.620 *
	Syringylacetone	0.612 *
	cis-Isoeugenol	0.591 *
	Arabinose	− 0.556 *
Virgin brightness	Holocellulose	0.795 **
	Arabinose	− 0.552 *
Heated brightness	Holocellulose	0.670 **
	Extractives	− 0.595 *

*** Significant at 1% level, **Significant at 5% level,
* Significant at 10% level

Screened pulp yield had a significant relationship with cellulose content positively (+) and with hemicellulose content negatively (−), but not with extractives content. This is possibly caused by low extractives content in *E. globulus* of about 3-4% compared with other species [13]. The CTMP process is maintained to retain chemical compounds as much as possible and the results obtained are reasonable because hemicellulose can be easily water-solubilized and lost during pulping. The hemicellulose composition was extracted as important traits as xylan (−), glucomannan (+) and arabinan (−). These hemicellulose compnents might represent pH difference for lignin biosynthesis environment, which was proposed to cause the structural difference of lignin with polysaccharides as glucomannan is more acidic compared with others and will give more conjugated lignin structures hardly digested in the CTMP process [14]. Further studies are expected. Many lignin fragments obtained by pyrolysis were also extracted. Almost all of positively related lignin fragments possessed benzene ring with a side-chain containing carbonyl group and/or C=C or a side-chain containing C=C with no functional group. They may be derived from not conjugated lignin structures inside the cell wall considered from chromatograms of dehydrogenatively synthesized lignin [15] and of native lignin [8]. Since they were for both S and G units, lignin S/G ratio was not extracted as an important trait. On the other hand, negatively related lignin fragments possessed benzene ring with no side-chain or a side-chain of propenyl alcohol for S unit only. They may be cleaved from the reducing end of lignin [8] and can be created by the CTMP process with sodium sulfite.

Virgin brightness had a significant relationship with holocellulose content positively (+) and with arabinan composition in hemicellulose negatively (−). However, in heated brightness, extractives content was extracted negatively. This may imply the residual extractives in CTMP. The result obtained agreed eith a previous report [16].

Consequently, the relationship between pulp and wood properties was significantly high and can be explained reasonably. Tree selection for high pulp yield, high virgin brightness and low color reversion by dry heat exposure can be achieved by the analysis of wood properties. Since holocellulose content had a significant relationship with all utilized traits positively, only holocellulose content analysis may allow the elite tree selection. Because the representative height of holocellulose content can be decided to take an increment core by using a radially divided increment core method [13], the elite tree can be retained for its future breeding, polination and propagation.

The feasibility of elite tree selection for effective utilization of forest resources toward the construction of efficient carbon loop was certified. Selected trees are expected to suppress CO_2 increase in the air, effectively.

References

[1] FAO. "The global forest resources assessment 2000", FAO (2001).
[2] Miranda, I., Almeida, M. and Pereira, H. For. Ecol. Manag. 149 (2001) 235.
[3] Grossmann, K. and Ott, G. Papier 48 (1994) 623.
[4] Ona, T., Sonoda, T., Shibata, M. and Fukazawa, K. Tappi J. 78 (1995) 121.
[5] Haygreen, J. G. and Bowyer, J. L. "Forest products and wood science. An introduction. 2nd Ed.", IOWA State University Press, Ames, IO (1989) 191.
[6] Ona, T., Sonoda, T., Ito K. and Shibata, M. Wood Sci. Technol. 31(1997) 205.
[7] Ona, T., Sonoda, T., Ito, K. and Shibata, M. Holzforschung 51 (1997) 396.
[8] Yokoi, H., Ishida, Y., Ohtani, H., Tsuge, S., Sonoda, T. and Ona, T. Analyst 124(1999) 669.
[9] Ona, T., Sonoda, T., Ito, K., Shibata, M., Ootake, Y., Ohshima, J., Yokota, S. and Yoshizawa, N. Appl. Spectrosc. 53 (1999) 1078.
[10] Ona, T., Sonoda, T., Ito, K., Shibata, M., Ootake, Y., Ohshima, J., Yokota, S. and Yoshizawa, N. Analyst, 124 (1999) 1477.
[11] Ito, K., Kato, T. and Ona, T. J. Raman Spectrosc. 32 (2001) 389.
[12] Efron, B. "The Jackknife: the boostrap and other resampling plans", Society for industrial and applied mathematics, Philadelphia, PA (1982).
[13] Ona, T., Sonoda, T., Ito, K., Shibata, M., Tamai, Y. and Kojima, Y. Appita J. 49 (1996) 325.
[14] Terashima, N. and Fukushima, K. "Plant cell wall polymers", Eds. Lewis, N. G. and Paice, M. G., ACS Symp. Ser., Vol. 399, Am. Chem. Soc., Washington DC (1989) 160-168.
[15] Sonoda, T., Ona, T., Yokoi, H., Ishida, Y., Ohtani, H. and Tsuge, S. Anal. Chem. 73 (2001) 5429.
[16] Ahmed, A., Kokta, B. V. and Carrasco, F. Cell. Chem. Technol. 29 (1995) 725.

Toward the Construction of an Efficient Link Between Forest Recycling and Paper Recycling Using Trees with High Performance for Paper Recycling

Toshihiro Ona[a], Jyunsuke Kawana[b], Yoko Kibatani[b], Yukiko Ishikura[c] Yasuo Kojima[d] and Takayuki Okayama[b]

[a]*Graduate School of Bioresource and Bioenvironmental Sciences, Kyushu University, Hakozaki, Higashi-ku, Fukuoka 812-8581, Japan*

[b]*Graduate School of Agriculture, Tokyo University of Agriculture and Technology, Fuchu 183-8509, Japan*

[c]*CREST of JST, Faculty of Agriculture, Tokyo University of Agriculture and Technology, Fuchu 183-8509, Japan*

[d]*Graduate School of Agriculture, Hokkaido University, N9-W9, Kita-ku, Sapporo 060-8589, Japan*

Introduction

After the Kyoto Protocol, considerable attentions have been paid on forests in the context of climate change. Forests have a key role in balancing society's need for wood-based products and protecting our environment by increasing net carbon sequestration in terrestrial carbon sinks removing atmospheric CO_2. The sustainably managed forests area is 187 million ha in 2000 and gains 4.5 million ha/year all over the world [1]. One of the forest management, the elite tree selection can improve forests in growth providing an effective way to remove more atmospheric CO_2. However, since carbon sequestration is at its best in growing young trees, recycling forests by harvesting and planting trees are essential.

From harvested trees, total 179 million tons of paper products per year were manufactured in 2001 [2]. The paper products have a new life as secondary raw materials after consumed and collected separately. Total 146 million tons of recovered papers were produced per year in 2001 and world recycling rate is 44.9% [2]. Paper recycling plays an important role to prevent land filling, which fatally emits one of greenhouse gases methane, to conserve natural resources and to reduce the amount of electricity used in paper making to reduce the greenhouse gas emission.

The mechanism of recycling paper is as follows [3]. Paper is formed mainly by hydrogen bonding between fibers. When water is removed during drying, significant surface tension happens between swollen or separated lamellae in fiber wall and between fibers, compacts and draws fibers into more rigid hydrogen bonding. When paper is dispersed into water, water swells fiber wall again and cleaves interfiber bonding into fiber-water-fiber structure to be each fiber.

However, repeated wetting and drying causes delamination, crack formation and shrinkage of fiber wall leading the wall surface inactivated to form weak hydrogen bonding between fibers, to deteriorate and to swell less in water. Consequently, fibers are not enough to be recycled into new paper products because of short and weak fibers. Therefore, the industrial recycling rate is limited to about three times and the smallest fibers are lost from the paper making process. This is similar to that we can wash a cotton shirt, mainly comprised of same cellulose as paper only three times. However, tree breeding program has never aimed for paper recycling.

The first purpose of our study here was to examine the feasibility of selection of trees with high performance for kraft paper recycling from plantation trees even showing similar growth and pulp properties after digestion in *Eucalyptus globulus* as a model tree, which is one of the most popular hardwood plantation species [4]. On the other hand, tree selection is preferably performed by a non-destructive way, which can retain the tree for future breeding including cross-pollination and propagation.

The second purpose was to develop a rapid and non-destructive method for elite tree selection aiming paper recycling together with an automatic instrument toward the construction of an efficient link between forest recycling and paper recycling.

Experimental

Materials

Total thirty two 9.5-year-old *Eucalyptus globulus* trees were obtained from Western Australia grown under the same silvicultural condition. The climate is a moderate zone with the mean annual maximum temperature of 20 °C and minimum of 10 °C. The mean annual rainfall is 1,069 mm and 850 mm. Chip samples representing each whole tree were prepared by cutting a log into pieces of approximately 20 mm width with 2 mm thickness.

KP

Kraft pulps were prepared in a digester with 400 g oven dry chip charges. Kraft pulps with a target Kappa number of 20±2 were prepared from each chip sample under various cooking conditiona at constant alkali charge. The cooking conditions were as follows: active alkali charge, 11-12%; sulphidity, 25%; liquor-to-wood ratio, 4; time to maximum cooking temperature, 170°C. Time at the maximum temperature was changed to obtain the target Kappa number. The pulps were screened by a 12-cut flat screen. Total pulp yield, rejects and Kappa number were determined described previously [5].

Paper recycling and measurement of tensile index of handsheets

Paper recycling and measurement of tensile index of handsheets were performed same as Ref [5].

Raman spectroscopy

Raman spectra were collected using a Nicolet Raman 960 spectrometer (Thermo Nicolet, Madison, WI, USA) equipped with an InGaAs detector, and an Nd: YAG laser operating at 1064 nm. Wood powder samples (less than 0.84 mm) were packed into nuclear magnetic resonance tubes and spectra were collected using 180° backscattering at 500 mW, 4 cm^{-1} resolution and 512 scans under 22°C using a thermal box [6]. Spectra of 100-3700 cm^{-1} Raman shift were used and second-derivatized after multiplicative scatter correction. The multivariate data analysis including the partial least squares (PLS) regression was performed by using the software of Unscrambler 7.5. Important frequencies were further selected by the Marten's uncertainty test based on Jackknife validation method to improve the calibration model [7, 8]. The number of factors used in the PLS calibration model was decided by full-cross-validation method.

Results and Discussion

Tree selection

The recyclability was assessed by non-linear regression of tensile index curve depending on recycling same as Red [5] and subsequent normalization of all samples to be the recyclability to be three times.

The tree propeties of *E. globulus* and elite tree candidates were shown in Table 1.

Table 1 Tree data.

Trait	Tree #1	Tree #2	Tree #3	Tree #4	Min	Average	Max
Volume (m^3)	0.438	0.090	0.351	0.473	0.082	0.256	0.473
Pulp yield ($kg\ m^{-3}$)	270	229	264	265	206	273	328
Recyclability (times)	6.76	7.22	6.13	7.12	0.64	3.00	7.22

Large between-tree variation was observed in tree volume under the bark and pulp yield although they were silvicultured and grown equivalently. In recycling, large between-tree variation was observed similar to volume and pulp yield as shown in Figure 1.

A selected elite tree provided 1.85 times higher volume (growth), 0.97 higher pulp yield and 2.37 times higher recyclablity compared with the average of utilized 31 individuals. The tree selection will help the construction of efficient carbon loop.

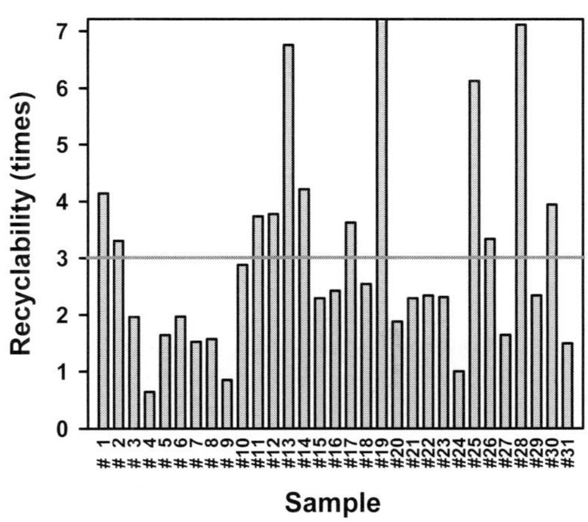

Figure 1 Between-tree variation of recyclability.

Rapid and non-destructive method for elite tree selection aiming paper recycling

Plots of calibration and validation in the PLS regression are shown in Figure 2a and 2b, respectively. The statistical results were; correlation coefficient (r): 0.996, root mean square error of calibration: 0.16 in calibration, and r: 0.918, root mean square error of prediction: 0.70 and number of used factors; 5. Used factors were decided by the predicted residual error sum square.

As a result, we successfully obtained highly significant correlation coefficients between measured and Raman predicted values. This was the first successful non-destructive quantitative calibration of recyclability using vibrational spectroscopy combined with multivariate data analysis.

The feasibility of elite tree selection for effective utilization of forest resources toward the construction of efficient carbon loop was certified. Selected trees are expected to suppress CO_2 increase in the air, effectively.

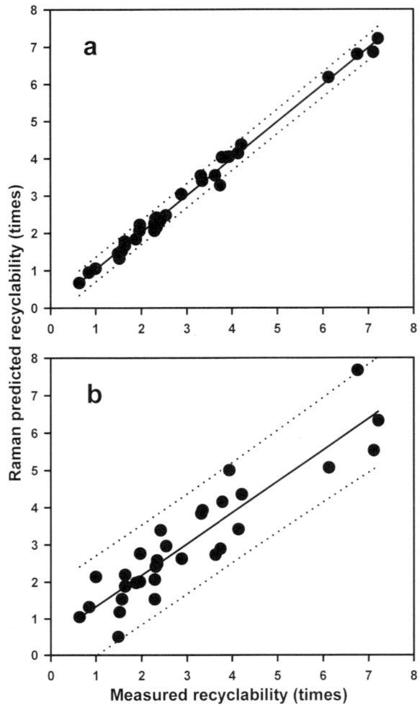

Figure 2 Plots of calibration (a) and prediction (b) for recyclability.

References

[1] FAO. "The global forest resources assessment 2000", FAO (2001).
[2] Annual review. Pulp Paper Int. 44 (2002).
[3] Emerson, H. W. "The preparation of pulp fibres for papermaking", in "Handbook of Paper Science, Vol. 1, The raw materials and processing of papermaking", Ed. Rance, H. F., Chapter 3, Elsevier, Amsterdam (1980).
[4] Clark, C. R. E. Appita J. 53 (2000) 231.
[5] Okayama, T., Kibatani, Y., Nakayama, K., Kojima, Y. and Ona, T. in this book.
[6] Ito, K., Kato, T. and Ona, T. J. Raman Spectrosc. 32 (2001) 389.
[7] Efron, B. "The Jackknife: the boostrap and other resampling plans", Society for industrial and applied mathematics, Philadelphia, PA (1982).
[8] Wold, S. Technometric 20 (1978) 397.

Chapter 2

Pulp and Paper Recycling

Chemical Aspects of Photo-Yellowing of High-Yield Pulps

Göran Gellerstedt and Liming Zhang
Department of Fibre and Polymer Technology, Royal Institute of Technology, KTH, S-100 44 Stockholm, Sweden

Introduction

High-yield pulps such as TMP and CTMP turn yellowish on exposure to daylight (photo-induced yellowing) or heat in the presence of moisture (thermal yellowing). Whereas the latter type of yellowing may have an importance in e.g. the production process and in the paper machine, the former type predominantly occurs as a non-desirable process when the finished paper product is exposed to light. It has also been shown that the yellowing becomes more prominent when the brightness is high [1].

The chemistry behind the photo-induced yellowing is complex but most studies to date indicate that the lignin portion present in mechanical pulps is mainly responsible for the photo-chemical reactions to occur [2]. In order to understand what chemical factors are of importance, a knowledge about the complete chemistry encountered when wood is converted into pulp and further bleached as well as the solid-state photo-chemistry of the various wood components must be acquired, however. In the present paper, the formation of photo-active groups during the pretreatment of wood, the refining and further bleaching of mechanical pulps will be discussed. With this as a background, the differences in yellowing tendency of different types of high-yield pulps will be exemplified together with suggestions of possible chemical reactions.

Results and Discussion

During the refining of wood, a certain formation of oxidized lignin structures takes place [3]. The most important chemical change seems to be the cleavage of a small number of β-O-4 structures and formation of new phenolic end-groups together with structures of the Hibbert ketone type. If the refining is preceded by a treatment with aqueous sodium sulfite (CTMP, CMP), the number of newly formed phenolic groups increases thus indicating a somewhat higher degree of cleavage. The content of phenolic hydroxyl groups is shown in Figure 1 for stone groundwood, bleached with hydrogen peroxide, and for some different types of CTMP and CMP. Wood has been used as a reference. An almost linear relationship between the number of phenolic hydroxyl groups and the severity of the treatment with sulfite was obtained. These results are interpreted as being due to a combination of β-O-4 cleavage caused by ionic reactions with sulfite and a homolytical cleavage due to the high temperature in the refiner.

In the bleaching of mechanical pulps to high brightness, alkaline hydrogen peroxide is employed and applied in such a way that the dissolution of wood substance is kept at a minimum. Thereby, a brightness maximum of around 80% ISO can be reached for spruce pulps. In the brightening reactions, various conjugated carbonyl structures present in the lignin are partly or completely oxidized by peroxide to form carboxyl groups but, to some extent, other structures such as hydroquinones can also form [4, 5]. These together with catechol structures, originating from the wood, may easily be converted into colored products by the action of light or heat [6, 7].

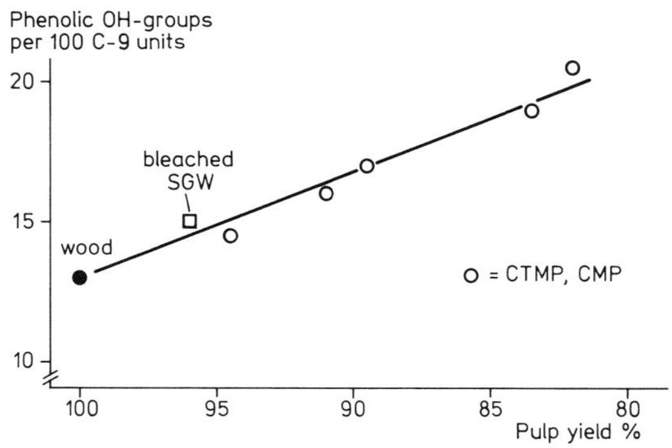

Figure 1 Content of phenolic hydroxyl groups in spruce wood and some mechanical pulps.

All types of mechanical pulps from spruce have a high degree of photo-induced yellowing but large differences exist between pulps produced in different types of processes. Thus, a relationship seems to exist between the severity of the chemical pretreatment and the degree of yellowing if the comparison is done for pulps having a similar brightness. This is shown in Figure 2 for stone groundwood, CTMP and CMP with the latter having the highest brightness reversion. Furthermore, the differences in yellowing between the various chemimechanical pulps were found to correlate well with the content of phenolic hydroxyl groups.

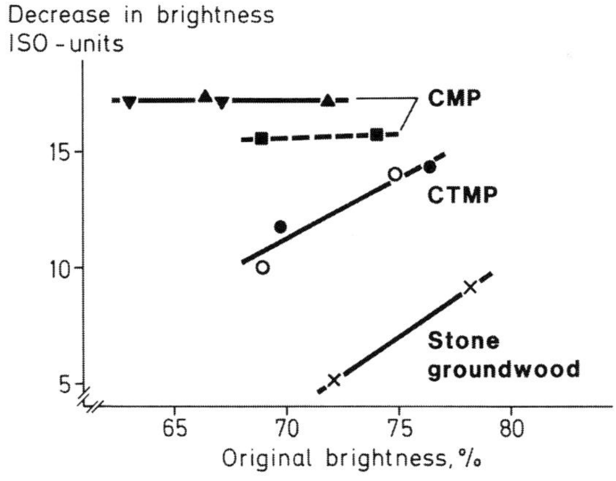

Figure 2 Photo-induced yellowing of various spruce mechanical pulps bleached to different brightness levels.

In order to get more information about the types of structures contributing to color as a result of irradiation, several quinone model compounds have been synthesized and their light absorption characteristics measured in the solid state. Thereby, it was found that largely different values for the light absorption maximum can be obtained depending on the type of solid matrix used. This is illustrated in Figure 3 where it is shown that filter paper and bleached groundwood pulp give rise to large differences in the light absorption curve of a simple o-quinone, absorbed on the paper. Thus, any attempt to get information about the structure of the chromophores being formed in the pulp as a result of irradiation must be done using the actual type of mechanical pulp. Some further differences in absorption characteristics for chromophoric compounds are shown in Figure 4 and Table 1.

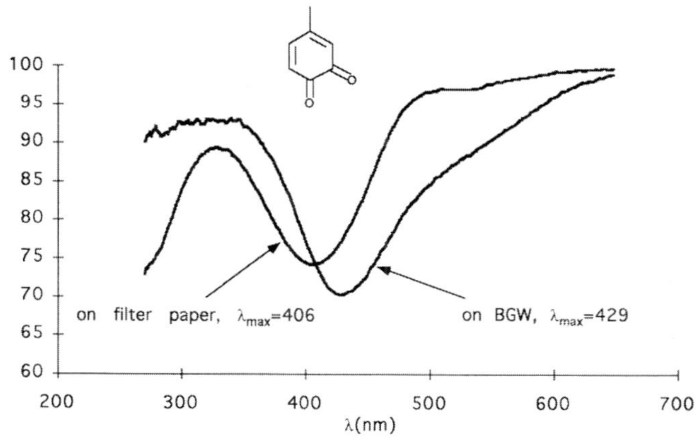

Figure 3 Light absorption maximum for 4-methyl-o-quinone, absorbed on either filter paper or bleached groundwood (BGW).

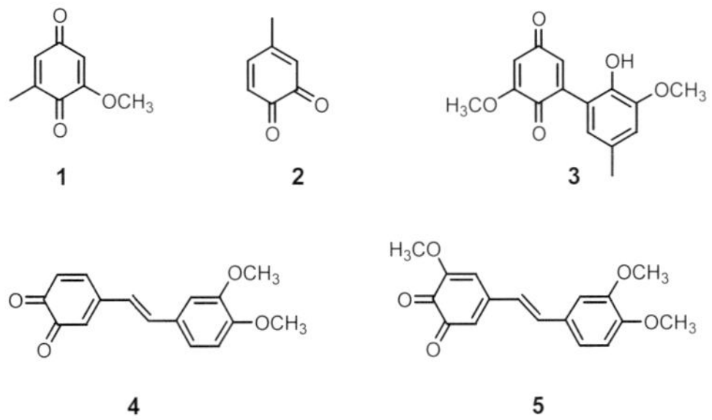

Figure 4 Model compounds used in the evaluation of light absorption of chromophores.

Table 1 Light absorption maxima for model compounds shown in Figure 4.

Medium	Model compound				
	1	2	3	4	5
Methanol	357	387	328	423	-
Filter paper	375	406	338	-	-
BGW*	416	429	-	-	-
BCTMP**	417	423	425	-	460

* BGW = Bleached spruce stone groundwood
** BCTMP = Bleached spruce chemithermomechanical pulp

In the mechanism, responsible for the photo-induced oxidation of mechanical pulps, it is assumed that an accessible hydrogen atom in a phenol or alcohol group is first abstracted by a light-excited functional group, e,g a carbonyl group in lignin. Thereby, a radical species is formed which in further steps is oxidized into a carbonyl group. From aromatic structures, the corresponding quinone is formed [6] but oxidation can also occur in other structural units present in lignin as well as in carbohydrates [8]. The possibility of carbohydrate oxidation has not been well documented but in a model experiment it could be shown that a strong discoloration can be obtained on irradiation if a lignin-like sensitizer is added to filter paper. Thus, addition of either acetoveratrone or vanillin gave rise to a strong brightness decrease despite the fact that acetoveratrone alone cannot form colored reaction products (Figure 5).

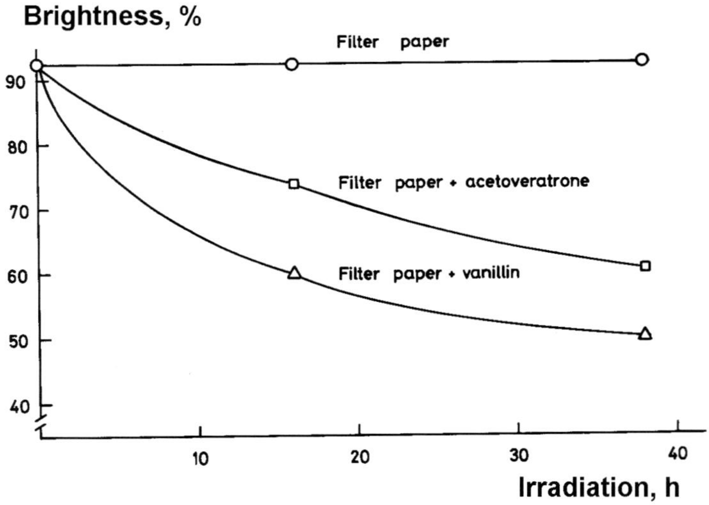

Figure 5 Photo-induced yellowing of filter paper impregnated with either acetoveratrone or vanillin.

References

[1] Gellerstedt, G., Pettersson, I. and Sundin, S. Svensk Papperstidn. 86 (1983) R157.
[2] Kringstad, K. Tappi J. 52 (1969) 1070.
[3] Johansson, M., Zhang, L. and Gellerstedt, G. Nordic Pulp Pap. Res. J. 17 (2002) 5.
[4] Gellerstedt, G. and Agnemo, R. Acta Chem. Scand. B34 (1980) 275.
[5] Gellerstedt, G., Hardell, H-L. and Lindfors, E-L. Acta Chem. Scand. B34 (1980) 669.
[6] Gellerstedt, G. and Pettersson, E-L. Svensk Papperstidn. 80 (1977) 15.
[7] Gellerstedt, G. and Pettersson, B. Svensk Papperstidn. 83 (1980) 314.
[8] Holmbom, B., Ekman, R. and Eckerman, C. Proceedings of 5th International Symposium on Wood and Pulping Chemistry, Raleigh, Vol. 1 (1989) 445-450.

Production of High Brightness CTMP from *Eucalyptus globulus* and Their Light-Induced Color Reversion

Yasuo Kojima[a], Shinichi Isaji[b] and Toshihiro Ona[c]
[a]*Graduate School of Agriculture, Hokkaido University, Sapporo 060-8589, Japan*
[b]*Hokkaido Forest Products Research Institute, Asahikawa 071-0198, Japan*
[c]*Graduate School of Bioresource and Bioenvironmental Science, Kyushu University, Fukuoka 812-8581, Japan*

Introduction

Many *Eucalyptus* plantations have been developed in the tropical and sub-tropical regions. *Eucalyptus globulus* is one of the fast-growing species planted in temperate regions and its normal rotation period is 8-10 years. It is suitable for kraft pulping because of its high pulp yield and the tree plantation is expected to solve the pulpwood supply shortage effectively in many regions in the world. However, more effective utilization of *E. globulus* trees as a raw material of pulp is demanded because kraft pulp yield is usually lower than mechanical pulp.

One of the mechanical pulps, chemithermomecahnical pulp (CTMP) is highlighted because of high yield with good strength. However, the CTMP has an issue of color reversion caused by light and/or heat exposures and the color stability should be required to obtain high quality pulp especially for paper recycling. The color reversion by light exposure is often caused by the combined two phenomena of light excitation and oxidation of lignin and extractives distributed in pulp fibers [1].

In this study, the variation of color stability by light and heat exposures among individuals are respectively examined form the concentration and structural features of lignin and extractives in the cell walls of the pulp fibers of *E. globulus* CTMP.

Experimental

Material

Eucalyptus globulus used in this study was supplied from CALM (Department of Conservation and Land Management, Western Australia). Those trees were prepared from two plantation sites (T stand and M stand) in Western Australia. Total four individuals from two sites were selected respectively, and the chips from those trees were prepared.

CTMP

Eucalyptus wood chips were pretreated with 5% sodium sulfite solution at 70°C for 1 hr and 20°C for 11 hr. First refining was conducted at 135°C, preheat; 5 min, defibration; 3min in a Defibrator type D. After first refining, PFI mil was used for second refining, pulp consistency; 20%, clearance; 0.5 mm. Refining pulps were screened by a laboratory flat screen with a 8 cut plate.

Color reversion

Hand sheets from CTMP were irradiated with UV light at 265 nm from 30 min to 60 min, brightness of sheets before and after treatment were measured. The PC numbers of sheets were calculated by using the brightness of sheets before and after heating at 105°C.

Changes of the UV spectra of phenols in the fiber of CTMP

The CTMP sheets irradiated with UV light were cut off (1 mm x 3 mm) and embedded in Epon. The cross sections (1 μm-thickness) were cut by a glass knife on an ultra-micro tome. Several cell walls in the sections were selected from irradiated side, not irradiated side and the between part of the paper. The UV spectra of the spot (1 μm-diameter) in the cell walls were collected.

Results and Discussion

The CTMP produced from several *E. globulus* trees from two plantation sites had high quality and these CTMP were obtained in high yields over 85%. Sodium sulfite should be one of suitable reagents as the pretreatment chemical for CTMP production with high brightness. While there are no difference in pulp quality between the populations of *E. globulus* trees from two plantation sites, large variation of pulp quality in a population from site T was observed (the brightness varied from 55.9 to 68.5). This variation in brightness should be caused by extractives [1]. In fact, pulp brightness was highly correlated to extractives of wood in this study (Figure 1). Therefore extractives content should be an important selection index of trees for the production of high brightness CTMP. Two individuals from the site T yielded high brightness CTMP over 67 (as ISO brightness) in high yield.

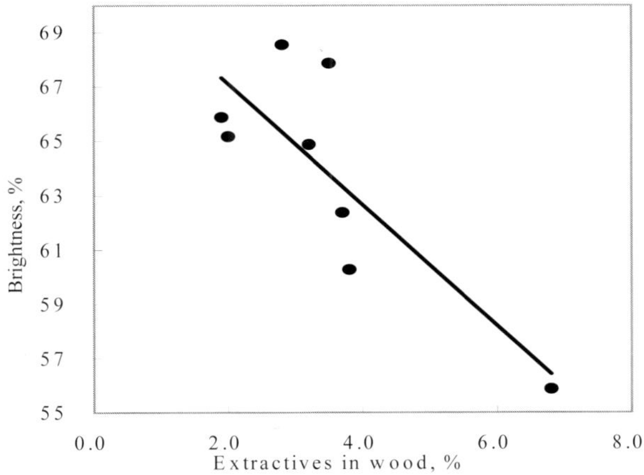

Figure 1 Correlation of pulp brightness with wood extractives.

Tensile indexes of CTMP from *E. globulus* trees were depended on the sheet density and CSF of pulps. In the second refining with PFI mill, same energy was consumed for production of each CTMP. Therefore, tensile indexes should be related to refining ability of pulp. These properties of CTMP were correlated to the chemical properties of eucalyptus wood. As shown in Figure 2, the sheet density as a factor of tensile strength was related to lignin content in correlation coefficient –0.654. Sheet density depended on the degree of sheet forming and lignin in the pulp fiber should inhibit the sheet formation. This should be reason why negative correlation of lignin content in wood with sheet formation was obtained.

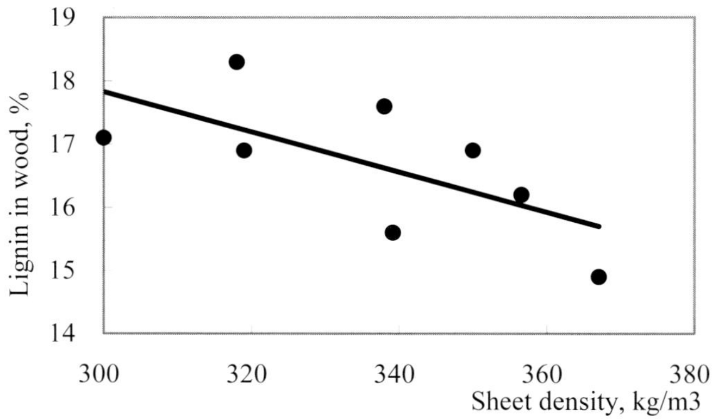

Figure 2 Correlation of pulp sheet density with lignin content in wood.

PC No., used as a factor of stability of brightness in bleached chemical pulp, was calculated from the brightness of CTMP before and after heating. Despite that the CTMP examined in this study contained a large amount of lignin and extractives, PC numbers of these CTMP were very low. PC number of CTMP from *E. globulus* trees in the plantation site M were under 2 points and these values are sufficient level in the bleached chemical pulps for the printing paper. It should be concluded that the PC number was due to the contents of unstable lignin in the pulp fiber. In high yield pulp, no chemical changes of the lignin structure in the pulp fiber were occurred compared with chemical pulps such as kraft pulp. PC number of these CTMP correlated to extractives of wood chips as shown in Figure 3.

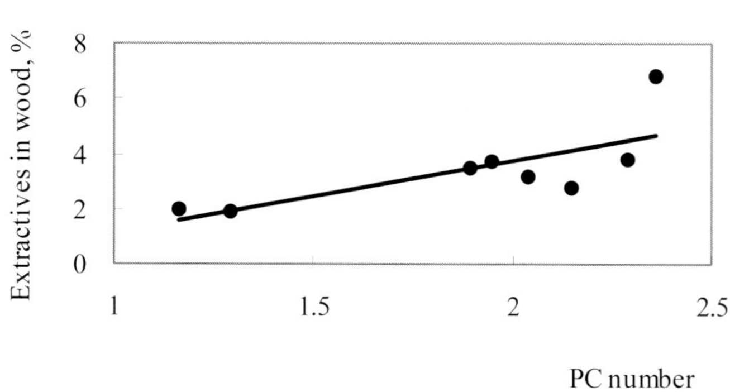

Figure 3 Correlation of PC number with extractives in wood.

In consider using these CTMP as raw materials for printing paper, it should be required to elucidate the stability of these CTMP to light at variable wave length as well as the heat stability. The results of the color reversion of CTMP by the irradiation of UV light at 265 nm are shown in Table1. Stability of the pulp brightness (⊿K60) to UV irradiation showed the negative correlation with extractives as shown in Figure 4. The correlation of extractives in wood with ⊿K60 was contrast to that with PC number. There are difference in the mechanism of the stability of brightness to heat and light. It is required that the more chemical information to understanding the mechanisms of the yellowing of these CTMP by the light irradiation. To elucidate the stability of these pulps to light, a new method was applied using a UV microscope spectrometer. The yellowing of pulp by light irradiation due to the chemical changes of phenols such as lignin and tannin in the pulp fiber and it should caused to the changes of UV spectra of the phenols in the cell wall of pulp fibers. The changes of the UV spectra of cell walls of CTMP fibers from T39 and M17 after light irradiation at 264 nm for 60 min are shown in Figures 5 and 6. In the case of CTMP from *E. globulus* M17, the UV absorbance in the cell wall of irradiation side in the region from 280 to 330 nm was higher than that in the cell wall of between and back side in the pulp sheets. The UV absorbance of irradiation side of CTMP from T39 was increased by UV irradiation in the region from 279 to 350 nm. These changes in the UV spectra indicated the formation of the conjugated structures in the cell wall of pulp fiber by UV irradiation and their structure should induce the yellowing to pulp.

Table 1 Change of CTMP brightness by UV irradiation.

	ISO Brightness(K) after UV irradiation				
	0min	10min	30min	60min	⊿K60*
M 1	65.16	60.54	56.23	52.09	20.02
M 4	65.86	61.48	56.58	52.76	19.89
M15	64.87	59.63	56.07	52.11	19.67
M17	62.37	58.15	54.84	50.81	18.53
T36	60.28	55.51	51.99	48.42	19.67
T37	55.87	52.67	48.78	46.24	17.24
T38	67.85	63.34	58.52	54.57	19.57
T39	68.54	61.64	57.57	53.20	22.38

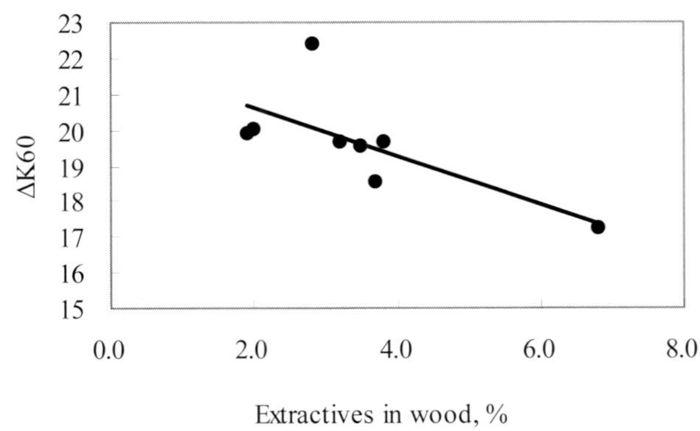

Figure 4 Correlation of ⊿K60 with extractives in wood.

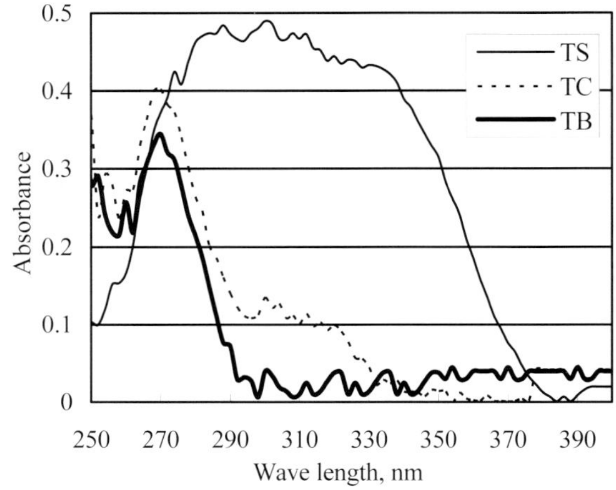

Figure 5 The changes of the UV spectra of T39. MS; irradiation side, MC; between two sides, MB; back side.

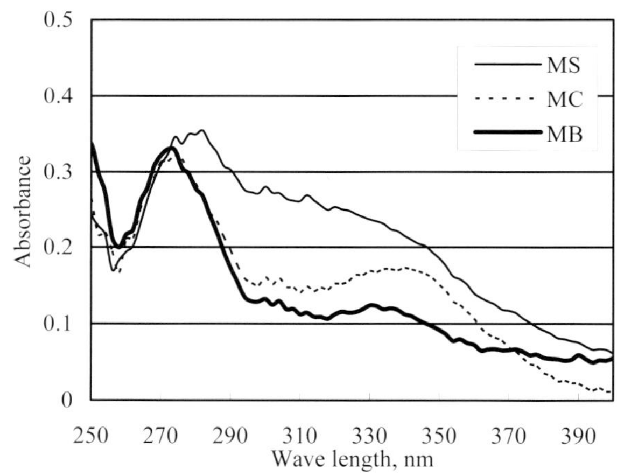

Figure 6 The changes of the UV spectra of M17. MS; irradiation side, MC; between two sides, MB; back side.

For further chemical information of yellowing mechanism of pulp, it is needed to obtain the more results of pulp yellowing under variable conditions.

In previous scheme, carbonyl group and phenolic group take a important role for photo-yellowing of papers caused by formation of quinones from lignin structures in the pulp fibers. However, new photo-yellowing mechanisms are proposed by new experimental fact and with new analytical methods. Mechanistic pathway that lead to the formation of stilbenes and/or o-quinones are not important [2], role of stilbenes in the photo- yellowing have been negative [3], strong support are provided for direct photo-oxidation of hydroquinone to p-quinone are experimentary showned, and hydroquinones were found to be present in pulps by Raman method [4].

We tried to irradiate the light to CTMP under variable condition and to evaluate the new mechanism described above. As shown in Figure 7, benzyl carbonyl groups are not important to produce the yellowing substances. This means the occurrence of direct elimination of proton radicals following by quinone structures. In fact, the same decreases of CTMP brightness are shown by irradiation under presence and absence of oxygen as shown in Figures 8 and 9. In the N_2, air and vacuum condition, irradiation of light on CTMP are yielded the same results, while brightness loss is small in O_2. This should be caused by radical scavenging by oxygen in reaction sites. Brightness loses of CTMP depend on the temperature under the light irradiation as shown in Figure 10.

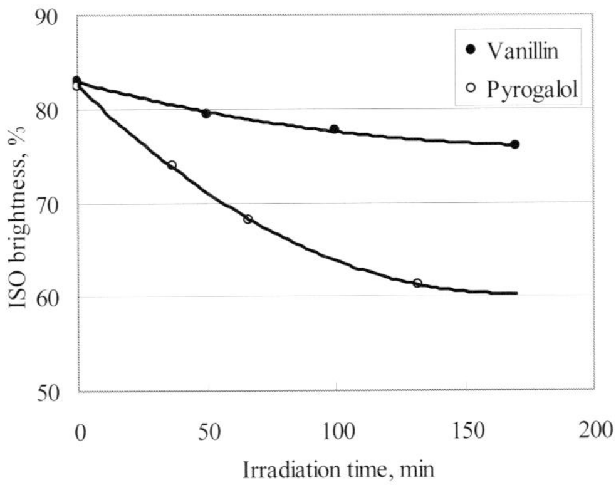

Figure 7 Photo-yellowing of vanillin and pyrogalol.

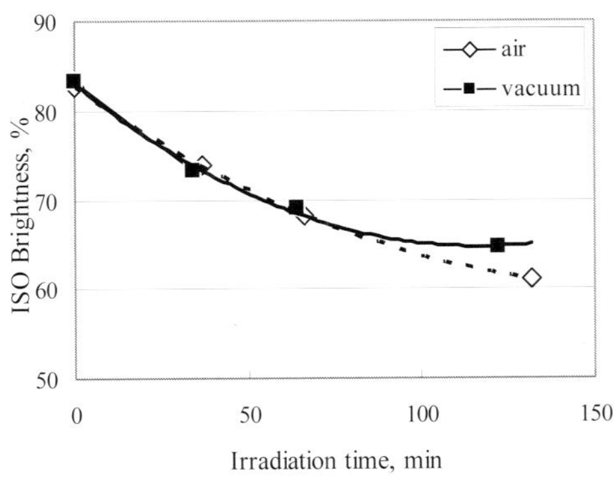

Figure 8 Brightness changes by irradiation with and without air.

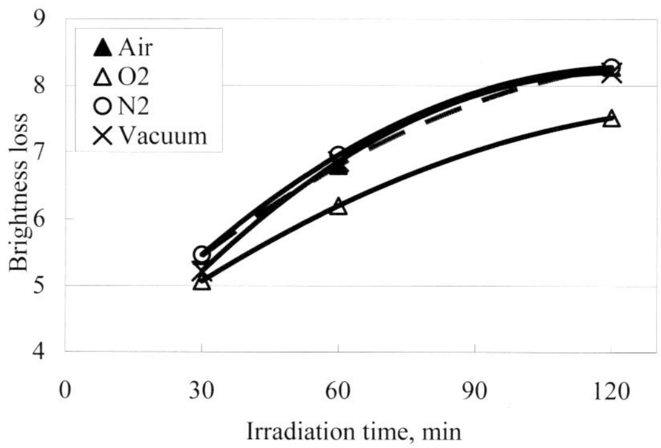

Figure 9 Effect of atmosphere on color reversion.

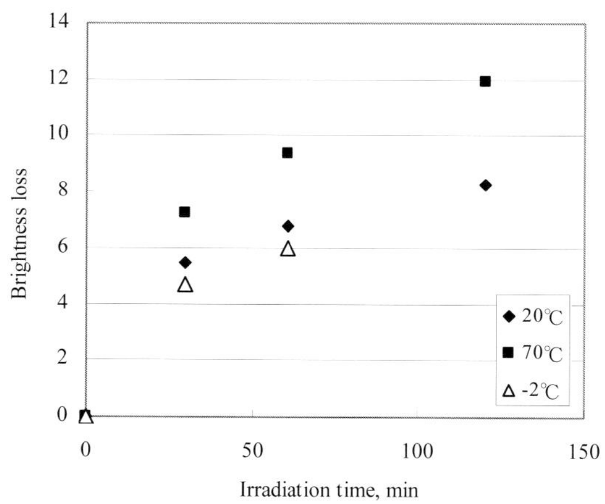

Figure 10 Effect of temperature on brightness loss.

References

[1] Ahmed, A., Kokta, B. V. and Carrasco, F. Cellulose Chem. Technol. 29 (1995) 725.
[2] Agarwal, U. P., Proceeding of 10th ISWPC, Vol. 1 (1999) 694-697.
[3] Ruffin, B. Can. J. Chem. 78 (2000) 73.
[4] Agarwal, U. P. J. Wood Chem. Technol. 18 (1998) 381.

Characterization of Photo-Yellowing Trigger Compounds Repressing Paper Recyclability of *Eucalyptus globulus* by Pyrolysis-GC/MS

Teruyuki Seino[a], Aki Yoshioka[b], Mitsuo Takai[b], Yasuo Kojima[c], Yukiko Ishikura[d], Toshihiro Ona[e], Yasuyuki Ishida[f], Hajime Ohtani[g] and Shin Tsuge[h]

[a]*NEDO Fellow, Institute for Environmental Management Technology, AIST Tsukuba West, Tsukuba 305-8569, Japan*
[b]*Graduate School of Engineering, Hokkaido University, Sapporo 060-8628, Japan*
[c]*Graduate School of Agriculture, Hokkaido University, Sapporo 060-8589, Japan*
[d]*Hokkaido Forest Products Research Institute, Asahikawa 071-0198, Japan*
[e]*CREST JST and Graduate School of Bioresource and Bioenvironmental Sciences, Kyushu University, Fukuoka 812-8581, Japan*
[f]*Research Center for Advanced Energy Conversion, Nagoya University, Nagoya 464-8603, Japan*
[g]*Graduate School of Engineering, Nagoya University, Nagoya 464-8603, Japan*
[h]*Graduate School of Engineering, Aichi Institute of Technology, Toyota 470-0392, Japan*

Introduction

Eucalyptus is a widely utilized species in tree plantation, especially intended for pulp production [1]. *E. globulus* is one of the fast growing species in temperate zone with a rotation period of 8-10 years [2]. We have reported that chemithermomechanical pulp (CTMP) paper prepared from several individuals of *E. globulus* showed high brightness (about 70% ISO) with more than 85% pulp yield [3]. To construct an effective carbon loop, higher paper recycling rate is preferable. The degree of photo-yellowing is crucial for recycling in CTMP paper because it determines requirement amount of bleaching agents. The photo-yellowing mechanism is proposed in which the photochemical reaction of lignin in pulp may lead to the formation of quinone species *via* intermediates of phenoxy radicals, which are produced through photoexcitation of the carbonyl groups involved in lignin (Scheme 1) [4].

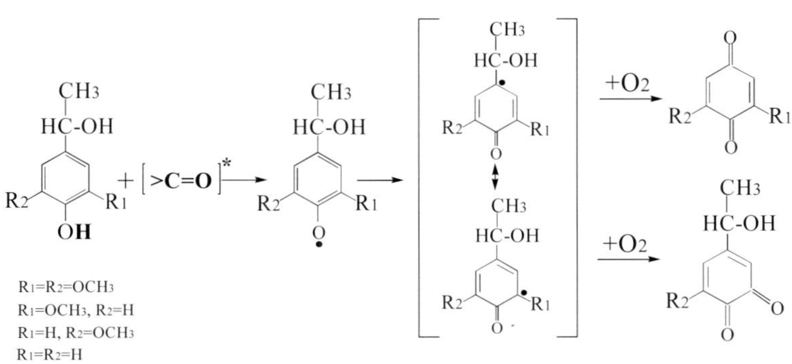

Scheme 1 Proposed mechanism of chromophore formation.

So far spectroscopic methods such as Fourier transform infrared (FTIR) [5,6], nuclear magnetic resonance (NMR) [7], and FT Raman spectroscopy [8] have been utilized to assign quinone species among the photo-induced compounds occurring upon photo-irradiation in isolated lignin and high yield pulps. Recently, Agarwal has assigned chromophore species produced in photo-yellowed pulp by using FT Raman and FT-IR spectroscopies. He suggested that such compounds would be formed from hydroquinones contained in original pulps [8]. However, these methods do not provide quantitative information because of the difficulty in discrimination of each individual species based on the observed complicated spectra.

On the other hand, pyrolysis-gas chromatography (Py-GC) has often been used for structural analysis of lignin [9-12] and pulp [13], which allows us their microgram-scale analysis without isolation of analyte from the sample matrices. This method has been also utilized for elucidating the degradation mechanism of various synthetic and natural polymers. Tsuge *et al.* investigated the photodegradative changes in chemical structure of silk by focusing on the changes in the sulfur-containing amino acid residues by means of Py-GC with sulfur chemiluminescence detection (Py-GC/SCD) [14]. Recently, we examined the difference in the amount of polyphenol components for CTMP paper samples prepared from *E. globulus*, which showed various yellowing levels after UV irradiation, by Py-GC in the presence of organic alkali, tetrabutylammonium hydroxide (TBAH). It was proved that the paper sample showing high yellowing after UV irradiation gave significantly large amount of phenol compounds containing a carbonyl group, known as photoexcitation species, during pyrolysis [15].

In this work, we propose a new method to analyze of quinone/hydroquinone species known as chromophores/leuco-chromophores derived during photo-irradiation in detail by Py-GC combined with mass spectrometry (MS) using CTMP paper samples.

Results and Discussion

Figure 1 shows the typical pyrograms of the CTMP paper sample prepared from *Eucalyptus globulus* (a) before and (b) after exposed to UV lamp obtained at 300°C. The relatively low temperature of 300°C was fairly controlled the peak intensities of pyrolyzates derived from major wood constituents. In addition to these peaks, on the pyrogram of the sample exposed to photo-irradiation [see Figure 1 (b)], quinone known as a chromophore was detected as a small peak at *ca.* 4 min. The mass spectrum for this peak is shown in Figure 2. In the spectrum, the molecular ion at *m/z* 108 was observed in addition to a fragment ion at *m/z* 80, mainly formed through the elimination of a -CO group. Moreover, the fragmentation pattern of this spectrum proved to be almost same as that of the standard spectrum of quinone. Therefore, the peak was assigned to quinone. Furthermore, hydroquinone was also a little observed on the program at *ca.* 15 min. The mass spectrum for this peak is shown in Figure 3. In the spectrum, the molecular ion at *m/z* 110 was observed in addition to a fragment ion at *m/z* 81, mainly formed through the elimination of a -COH group. Moreover, the fragmentation pattern of this spectrum proved to be almost same as that of the standard spectrum of hydroquinone. Therefore, the peak was assigned to hydroquinone. Since observed quinone species had never been separately detected directly by using the analytical methods including Py-GC, the results obtained are unique.

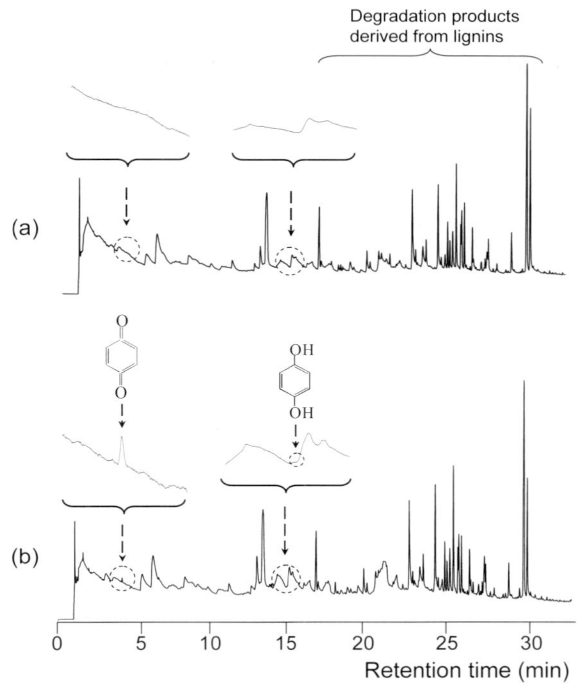

Figure 1 Pyrograms of CTMP papers from Eucalyptus globulus exposed to UV lamp obtained at 300°C by Py-GC/MS: (a) not exposed, (b) exposed to UV lamp.

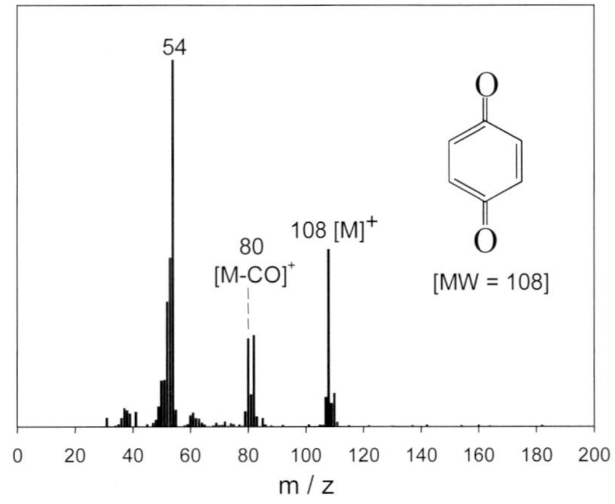

Figure 2 The mass spectrum of quinone observed in Figure 1.

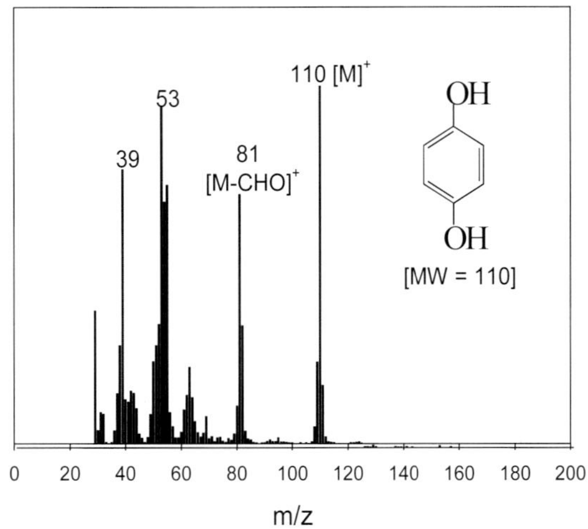

Figure 3 The mass spectrum of hydroquinone observed in Figure 1.

Recently, Agarwal has assigned chromophore species produced in photo-yellowed pulp by using FT Raman and FT-IR spectroscopies. He suggested that such compounds would be formed from hydroquinones contained in original pulps [8]. However, our results suggests that the hydroquinone species are not derived from that contained in the original pulps, as was proposed by Agarwal, but formed through the photo-derivation from lignin. Although it has reported that the hydroquinone type of lignin component without any methoxyl groups, so-called *para*-hydroxyphenylpropane (H) unit is contained at the level of 2~3 mol% in *E. globulus* tree [16], hydroquinone might be also derived from other origins. Leary has proposed that demethoxylation of guaiacyl lignin units in the photo-yellowing newspaper could result in the production of quinone or other colored products [17]. From this reports, hydroquinone could be considered to produce through the loss of methoxyl group of lignin during photo-irradiation. Therefore, the hydroquinone derived by the photo-irradiation was suggested to produce as intermediate species of chromophore.

Based on the results mentioned above, the probable mechanism for CTMP paper sample during photo-irradiation is proposed in Scheme 2. The photo-irradiated CTMP paper sample might proceed to form hydroquinone without any methoxyl groups which produced through the loss of those connecting to guaiacyl lignin mainly so that such compound may be caused the conversion to quinone with the rate of yellowing of pulp surface.

Consequently, the observed result suggested that the photo-yellowing of CTMP paper was closely related to the formation of quinone/hydroquinone without any methoxyl groups which was produced during photo-degradation of lignin structure.

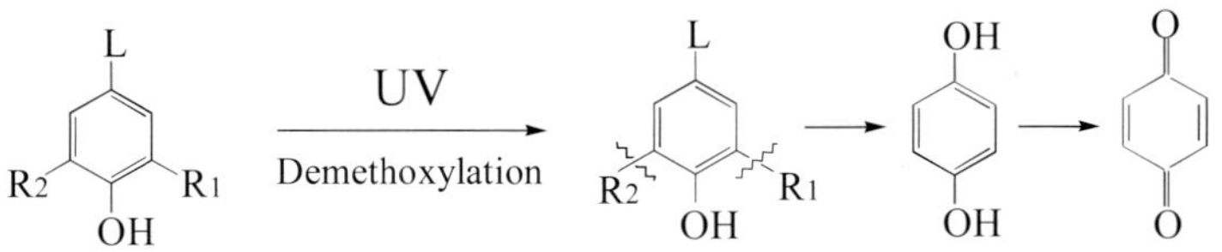

R1=R2=OCH3
R1=OCH3, R2=H
R1=H, R2=OCH3
R1=R2=H
L=lignin

Scheme 2 Possible chromophore formation pathway of CTMP papers.

References

[1] Shibata, M., Seni Gakkaishi 50 (1994) 141.
[2] Fao. Fao Forestry Series 11 (1981) 304.
[3] Kojima, Y., Isaji, S., Yoon, S. L. and Ona, T., 11th ISWPC, Nice, France, June 11-14, 2001; Vol. II, 215.
[4] Gierer, J. and Lin, S. Y. Sven. Papperstidn. 75 (1972) 233.
[5] Hon, D. N. S. and Chang, S. T. J. Poly. Sci. 22 (1984) 2227.
[6] Hon, D. N. S., Clemson, S. C. and Feist, W. C. Wood Sci. Tech. 20 (1986) 169.
[7] Argyropoulos, D. S. and Sun, Y. Photochem. PhotoBiol. 64 (1996) 510.
[8] Agarwal, U. P. J. Wood Chem. Tech. 18 (1998) 381.
[9] Yokoi, H., Ishida, Y., Ohtani, H., Tsuge, S., Sonoda, T. and Ona, T. Analyst 124 (1999) 669.
[10] Yokoi, H., Nakase, T., Ishida, Y., Ohtani, H., Tsuge, S., Sonoda, T. and Ona, T. J. Anal. Appl. Pyrolysis 57 (2001) 145.
[11] Sonoda, T., Ona, T., Yokoi, H., Ishida, Y., Ohtani, H. and Tsuge, S. Anal. Chem. 73 (2001) 5429.
[12] Greenwood, P. F., Van Heemst, J. D. H., Guthrie, E. A. and Hatcher, P. G. J. Anal. Appl. Pyrolysis 62 (2002) 365.
[13] Del Río, J. C., Iérrez, A., Romero, J. Martínez, M. J. and Martínez, A. T. J. Anal. Appl. Pyrolysis 58-59 (2001) 425.
[14] Tsuge, S., Yokoi, H., Ishida, Y., Ohtani, H. and Becker, M. A. Polymer Deg. Stab. 69 (2000) 223.
[15] Yoshioka, A., Takai, M., Kojima, Y., Seino, T., Ishikura, Y., Sonoda, T., Ona, T., Goto, K., Yokoi, H., Ishida, Y., Ohtani, H. and Tsuge, S. Pyrolysis 2002, Leoben, Austria, September 17-20 (2002) 129.
[16] Evtuguin, D. V., Neto, C. P., Silva, A. M. S., Domingues, P. M., Amado, F. M. L., Robert, D. and Faix, O. J. Agri. Food Chem. 49 (2001) 4252.
[17] Leary, G. J. Tappi J. 50 (1967) 17.

Fiber Properties and Papermaking Potential of Recycled *Acacia* Pulp

Takayuki Okayama[a], Yoko Kibatani[a], Kohei Nakayama[a], Yasuo Kojima[b] and Toshihiro Ona[c]

[a]*Graduate School of Agriculture, Tokyo University of Agriculture and Technology, 3-5-8 Saiwai-cho, Fuchu, Tokyo 183-8509, Japan*

[b]*Graduate School of Agriculture, Hokkaido University, N9-W9, Kita-ku, Sapporo 060-8628, Japan*

[c]*Graduate School of Bioresource and Bioenvironmental Sciences, Kyushu University, Hakozaki, Higashi-ku, Fukuoka 812-8581, Japan*

Introduction

The principle of paper recycling is based on that the inter-fiber bonding can be broken by hydration to isolate individual pulp fiber; that is, the sheet forming using water is a reversible process. It is interrogative whether every aspect in recycling is reversible, preserving the level of paper-making properties. It is obvious that most of paper properties in strength are lowered by the re-use of recycled chemical pulp fibers.

Although the recovery rate of wastepaper in Japan has exceeded 62% in 2001, the utilization rate of wastepaper in Japan was 58% and was lower than the recovery rate. Most of recovered papers have been utilized for lower-quality grade papers and paperboards, namely newspapers, toilet roll and corrugated board. For example, the average recycled fiber content is 58% for newspapers and more than 90% for corrugated boards. The utilization of recycled fibers for newspapers has been restricted in the last decade because of the decreasing demand of density (grammage). The utilization for printing and information papers is still at low level. Recent rapid increase of office papers consumption became to a serious problem in the waste management for local governments. The re-use of office papers for printing and communication papers has been limited due to concerns of paper appearance, strength and brightness. In order to promote the use of recovered papers for higher-quality grades, it is necessary to understand the fundamental aspects of recycled fibers.

Acacia mangium (AM) and *Acacia auriculiformis* (AA) are widely planted fast-growing hardwood species in tropical regions such as Malaysia and Indonesia. These acacias are potentially an important resource of pulpwood. AM has shorter and wider fibers than traditional fast-growing hardwoods such as *Eucalyptus grandis* [1]. Recently, some fundamental studies on pulping fast-growing AM have been carried out [2, 3].

On the other hand, the cross-pollination is being considered as a tree breeding method for the improvement of wood quality. The advantage of cross-pollination between AM and AA is described in this paper from fiber characteristics desirable for paper-making since few comprehensive studies have been reported for paper-making of *Acacia hybrid*.

This paper also covers the variations of several pulp fiber properties among trees and the relationships between fiber anatomical features and handsheet properties during recycling. Little published work is available on changes in the paper quality of fast-growing *hybrid* during recycling.

Experimental

Samples

Three species of 11-year-old fast-growing *Acacia* woods, namely *Acacia mangium*, *Acacia auriculiformis* and *Acacia mangium* x *Acacia auriculiformis* (*hybrid*), planted in Malaysia were randomly sampled, respectively. Total six trees considered to be representative in each species were debarked and chipped for kraft pulping. The chips representing a whole- tree were prepared by cutting the roundwood billet samples into pieces of approximately 20 mm-width with 2 mm-thickness.

Pulping

Kraft cooking was conducted in a digester with about 400 g oven dry chip charge. Kraft pulp with Kappa number of 15-23 was prepared from each chip sample by varying the cooking condition at constant alkali charge. The following cooking conditions were used: active alkali charge; 18%, sulfidity; 25%, liquor-to-wood ratio 3.5 and time to maximum cooking temperature; 170°C. Time at the maximum temperature was varied in order to attain the required Kappa number. The pulps were screened through a 12-cut flat screen. Total yield, rejects and Kappa number were determined after dewatering.

Preparation of recycled pulp fibers

Acacia unbleached kraft pulps were beaten in a PFI mill at 10% stock concentration for 7,500 rev. In order to prepare recycled pulps, 120 g/m^2 handsheets were prepared from each pulp sample based on the conventional standard method. The handsheets were dried at 80°C for 24 hours in a forced air circulation oven after the pressing. The dried handsheets were immersed into water for 1 hour and defibrated by a standard disintegrator. This was counted as one cycle and repeated up to five times.

Evaluation of pulps and handsheets

Canadian Standard Freeness, ISO brightness and water retention value (WRV) of the pulps were measured based on the ISO method and the JAPAN TAPPI pulp and paper testing method No.26. Length weighted averages in fiber length and fiber dimension such as width, wall thickness and lumen diameter were determined by a Kajjaani FiberLab. The ratio of lumen diameter/fiber width was calculated as indices of the shrinkage and collapse potentials of the dried and rewetted fibers. Physical evaluations were achieved with 60 g/m^2 handsheets prepared based on the ISO method 5269-1. Physical properties of the handsheet, namely sheet density, tensile strength, tearing strength, burst strength, folding endurance, light scattering coefficient and air permeance were determined.

Results and Discussion

Effect of recycling on kraft fiber property variation and on handsheet properties of Acacia species

Total 23 pulps prepared from individual trees, which were 6 *Acacia mangium* (AM), 7 AA and 10 *Acacia hybrid* (AH) trees, showed considerable between-tree variation within a species under freeness ranging from 325 ml CSF to 510 ml CSF at 7,500 PFI rev. The freeness of AM was lower than AA and the AH at the given PFI rev. The freeness of almost all pulp samples showed a slight increase with recycling. However, as for the several pulps, the freeness tended to decrease as the number of recycles increased though it once increased at the first cycle of wetting and drying.

Figure 1 shows the results concerning the mean tensile strength of the handsheets after five-times recycles. The variations in tensile strength of handsheets of the AH during recycling were similar to those in AM. However, handsheets from AA had lower tensile strength after recycling than other two *Acacia* species.

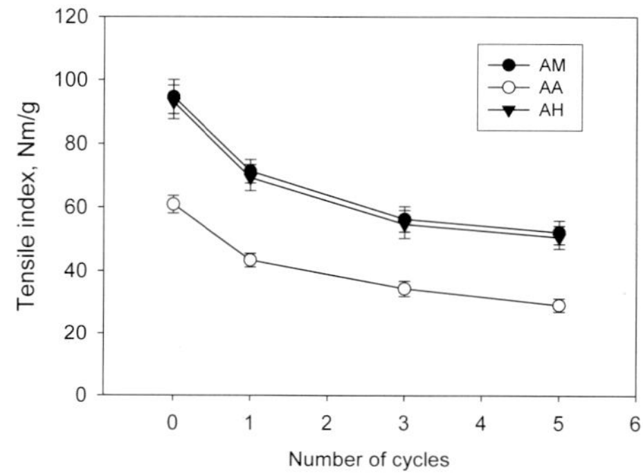

Figure 1 Effect of recycling on tensile index of Acacia handsheets.

The tensile strength is controlled by the strength both in individual fiber itself and interfiber bonding[4]. The zero-span tensile strength is primarily related to the individual fiber as distinguished from interfiber bonding. Figure 2 shows the mean zero-span tensile strength of *Acacia* handsheets. The zero-span tensile indices were almost constant during recycling. The results imply that the change in tensile strength of handsheet during recycling is due to a less extent to decrease in the individual fiber strength.

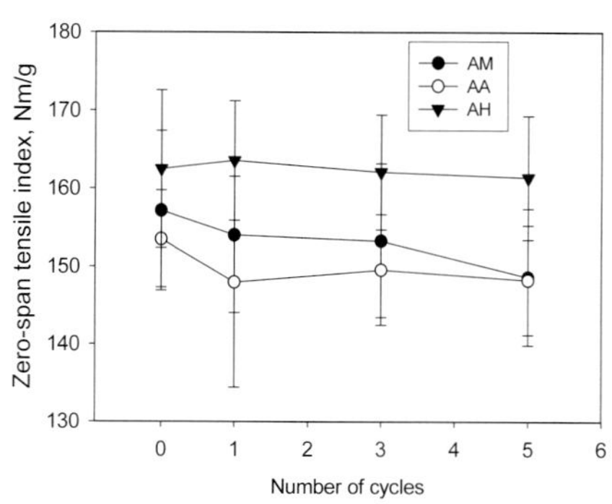

Figure 2 Effect of recycling on zero-span tensile index of Acacia handsheets.

Light scattering coefficient of *Acacia* handsheets increased with recycling. The light scattering coefficient of the AH during recycling was similar to that of AA (Figure 3). The light scattering coefficient closely related to the bonded area between pulp fibers. The loss in interfiber bonding during recycling was more remarkable than the change in intrinsic individual fiber strength.

The interfiber bonding is affected by fiber swelling which determines conformability of fibers. Figure 4 shows water retention values (WRV) of these pulp samples. AA pulps will be less conformable than AM and the AH because of the lower WRV. A rapid decrease was observed with number of cycles, particularly after the first cycle.

The kraft fiber property variation among trees of *Eucalyptus nitens* and *Eucalyptus fastigata* have recently been investigated [5-7]. Within-tree variation of fiber coarseness of *Acacia mearnsii* was compared with *Eucalyptus grandis* by a Kajaani FS 200 fiber analyzer [1].

Length-weighted mean fiber length was determined by a Kajaani FiberLab according to ISO 16065-1. A change was hardly observed in the fiber length during recycling (Figure 5). The fiber length variation was not high among three pulp samples. It could be expected that the tensile strength variation during recycling was not influenced by the pulp fiber length. It has been widely held concept for softwood pulps that tensile strength of paper is associated with fiber length. For hardwood pulps, fiber characteristics, such as collapse resistance (width/thickness), fiber perimeter and wall thickness or wall area combinations, and relative number of fibers, is also important for assessing paper qualities [8]. Figure 6 shows the results of fiber wall thickness variations during recycling. Wet fiber dimensions of wall thickness and width were also determined by a Kajaani FiberLab. The lumen diameter was calculated from fiber wall thickness and fiber width. The ratio of lumen diameter to fiber width was determined as the flexibility ratio of pulp fibers.

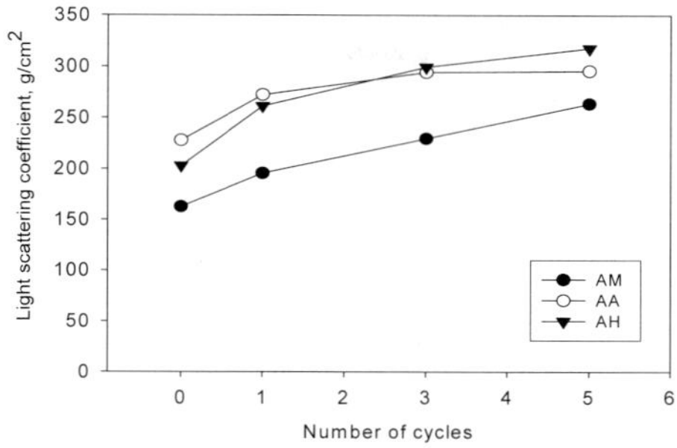

Figure 3 Effect of recycling on scattering coefficient of Acacia handsheets.

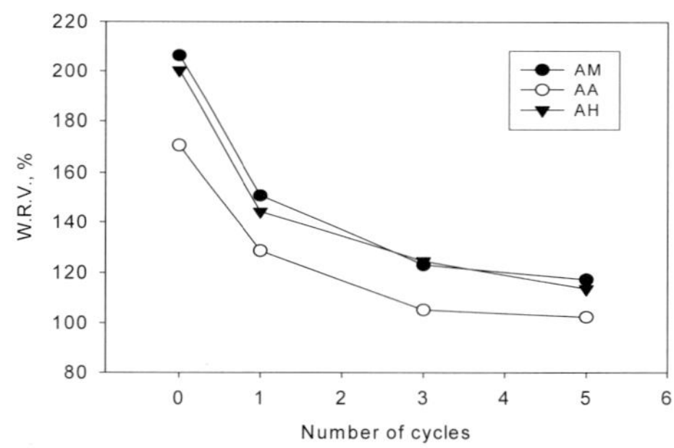

Figure 4 Effect of recycling on water retention value of Acacia pulps.

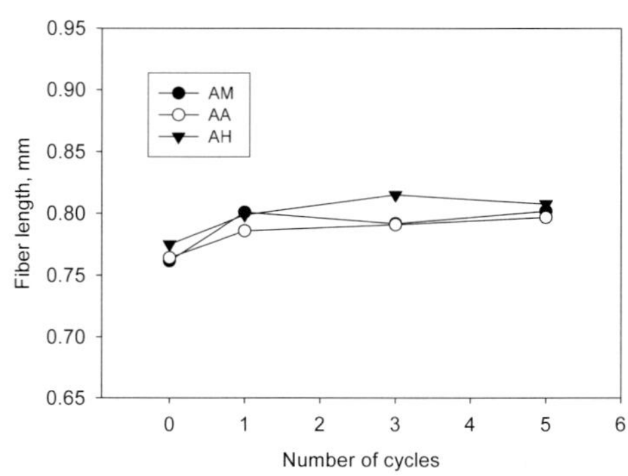

Figure 5 Effect of recycling on length-weighted average fiber length of Acacia pulps.

The fiber wall thickness, as shown in Figure 6, and fiber width of *Acacia* pulps decreased with increase in number of cycles. In particular, the more rapid decrease was shown in fiber wall thickness after the first cycle. On the other hand, lumen diameter of *Acacia* pulps showed the reverse tendency.

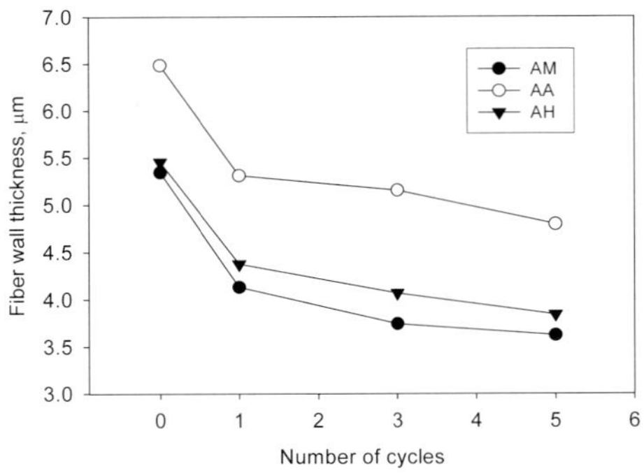

Figure 6 Effect of recycling on fiber wall thickness of Acacia pulps.

Figure 7 shows flexibility ratio (l/D) of three pulp samples. The flexibility ratio of pulps from AM and AH were higher than that of AA. The flexibility ratio increased with increasing number of cycles. It is widely recognized that the tensile strength of paper closely related to the flexibility ratio. Usually, the higher the flexibility ratio, the higher the tensile strength is.

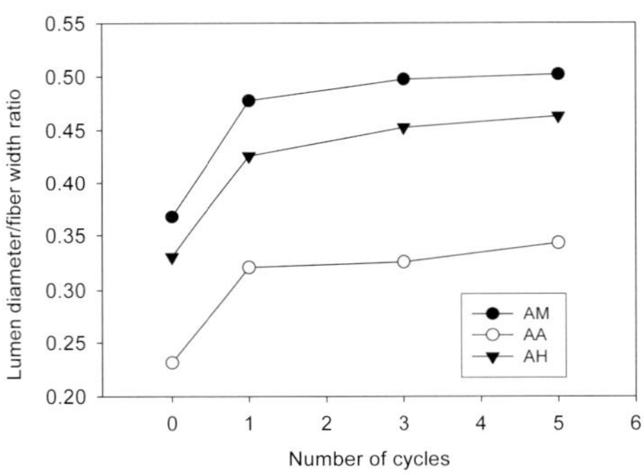

Figure 7 Effect of recycling on lumen diameter/fiber width ratio of Acacia pulps.

In previous papers, it was confirmed by the TEM micrographic observation of a cross section of the kraft pulp fiber that parts of the outer layer, S_1, of the pulp fiber were gradually separated from the fiber surface during recycling [9]. This is considered due to the difference in the shrinkage rates between S_1 and S_2 layers during recycling. The more parts of the fibrillated S_1 layer are removed from the fiber surface, the more the S_2 layer is exposed. The exposed S_2 layer is less accessible. Crevices form even in the radial direction of fiber cell walls with repeated recycling. Increases in the crystalline index by X-ray diffraction [10] and the pore volume in the cell walls of the pulp fiber [11] were also observed for recycled fiber.

When a pulp fiber dries, internal pore structure of the fiber wall is generated by stock preparation such as beating collapses. Although the potential for interfiber bonding is reduced, strong hydrogen bonding are formed by the capillary force within the fiber wall during recycling. It is not easy to reverse the collapse of internal structure in the fiber wall. It has been clarified that the shrinkage action in the fiber wall during recycling occurs perpendicular to the fiber axis rather than along the fiber axis.

The flexibility ratio of AA, which had lower tensile strength of handsheet after 5 cycles, was obviously lower compared to AM and AH. This is considered that the collapse in cross section of rewetted fibers must decrease as the flexibility ratio decreases at a given number of cycles. Pulp fibers with large width and thin wall thickness collapse to flattened fibers in the papermaking process to form a compact sheet with effective interfiber contacts.

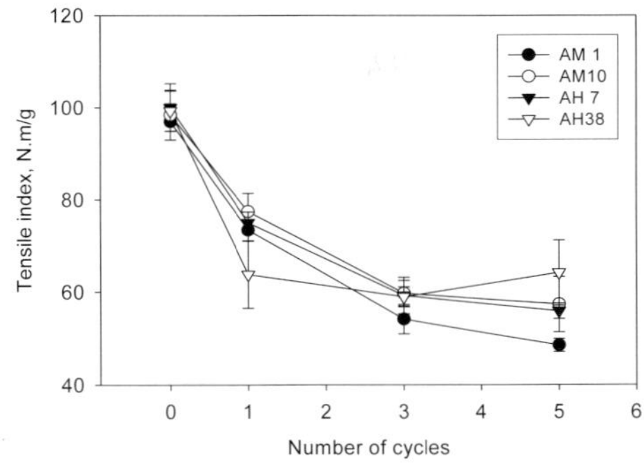

Figure 8 Effect of recycling on tensile index of handsheet from individual Acacia tree.

The pulps of AM-1, AM-10, AH-7 and AH-38 have equivalent initial tensile index at 7,500 PFI rev (Figure 8). Individual curve was generated for each sample. Pulp AH-38 was of high tensile index of the handsheet compared with other *Acacia* pulps after five cycles. On the other hand, pulp AM-1 gave the lowest tensile index after five cycles. As the pulp AM-1 showed a higher tensile index than pulp AM-10 after five cycles, there were significant tensile strength variations during recycling between trees and between species.

Prediction of tensile strength of handsheet at extended recycling

For all *Acacia* pulps, highly significant correlations were obtained between tensile index and number of cycles using the exponential regression model. An exponential decrease in tensile index was observed during recycling. We applied the exponential regression line for each *Acacia* pulp to calculate an expected tensile index at extended recycling. This made no significant difference to quality of the fit (R^2>0.986). The *hybrid* had higher expected tensile index, compared with other two species (Figure 9).

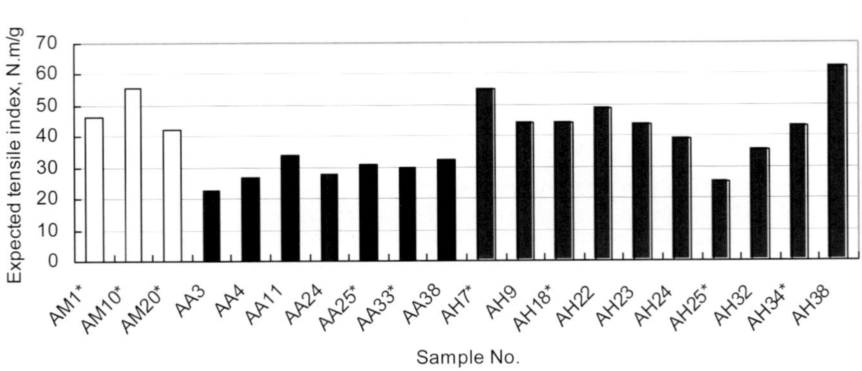

*Figure 9 Prediction of tensile index at extended recycling. (*significant at 0.05 level).*

Figure 10 shows that there was a correlation between sheet density and expected tensile index at extended recycling for each recycle number from three different *Acacia* handsheets. A better correlation was obtained for five times cycled *Acacia* pulps than for original pulps. The obtained correlation coefficients increased with number of cycles. Also the WRV of the original pulp showed highly significant correlation with the expected tensile index at extended recycling (Figure 11). As the WRV of the original pulp was over 200%, expected tensile index of >50 N.m/g was achieved at extended recycling. On the other hand, at the WRV of the original pulp less than 180%, the *Acacia* pulp gave expected tensile index of <30 N.m/g.

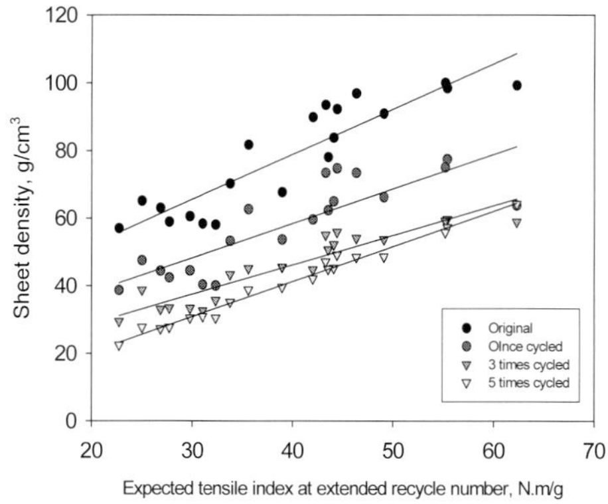

Figure 10 Relationship between sheet density and expected tensile index of Acacia handsheets.

The thinner the fiber wall thickness, the higher was expected tensile index at extended recycling (Figure12).

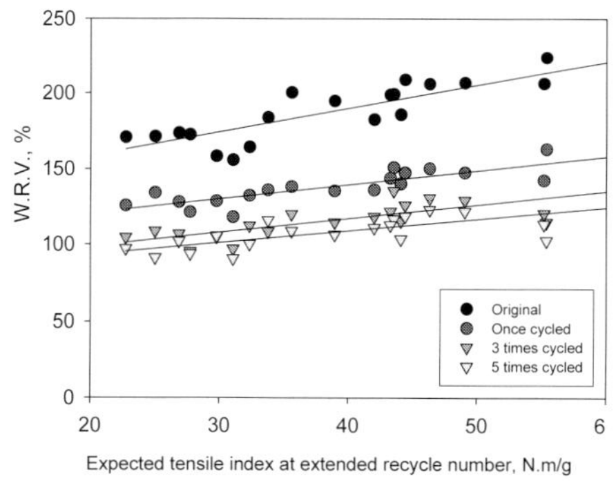

Figure 11 Relationship between water retention value of Acacia pulps and expected tensile index of their handsheets.

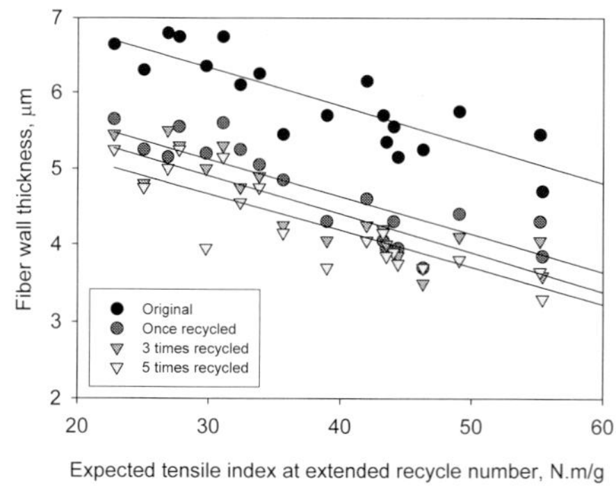

Figure 12 Relationship between fiber wall thickness of Acacia pulps and expected tensile index of their handsheets.

Conclusions

When kraft pulp is repeatedly wetted in water and dried for five times, the strength properties of handsheets are considerably deteriorated. Variations in fiber and handsheet properties of the twenty-three individual *Acacia* trees during recycling were investigated.

(1) A significant between-tree difference in the twenty-three individual tree pulps from *Acacia mangium, Acacia auriculiformis and Acacia mangium* x *Acacia auriculiformis* (*hybrid*) was observed for variation in tensile strength of handsheet during recycling.

(2) Tensile index of handsheet at extended recycling can be predicted by the exponential regression analysis for each *Acacia* pulp. Among the *Acacia* species in this experiment, the *hybrid*, which gave higher expected tensile index at extended recycling, appeared to be good for the application to paper-making, while *Acacia auriculiformis* was the least favorable.

(3) Expected tensile strength of handsheet at extended recycling closely related to the sheet density after five times cycles, the fiber swelling of the original pulp and fiber morphological characteristics. The increase in lumen diameter, as well as the decrease in fiber wall thickness and fiber width, was observed as a result of shrinkage configuration of the fiber wall during recycling.

References

[1] Muneri, A. and Balodis, V. Appita J. 50 (2000) 405.
[2] Xue, G.-X., Zheng, J.-W., Matsumoto, Y. and Meshitsuka, G. Japan Tappi J. 55 (2001) 366.
[3] Xu, E. C. Appita J. 52 (1999) 121.
[4] Page, D. H. Tappi J. 52 (1969) 674.
[5] Kibblewhite, R. P., Riddell, M. J. C. and Shelbourne, C. J. A. Appita J. 51 (1998) 114.
[6] Kibblewhite, R. P. and McKenzie, C. J. Appita J. 52 (1999) 218.
[7] Kibblewhite, R. P. and Riddell, M. J. C. Appita J. 53 (2000) 237.
[8] Kibblewhite, R. P. Appita J. 52 (1999) 429.
[9] Okayama, T., Yamagishi, Y. and Oye, R. Japan Tappi J. 36 (1982) 311.
[10] Yamagishi, Y. and Oye, R. Japan Tappi J. 35 (1981) 787.
[11] Okayama, T., Kitayama, T. and Oye, R. Japan Tappi J. 35 (1981) 1057.

The Impact of Chemical Composition of Pulp Fiber Cell Wall on Paper Recycling Potential of Fibers

Nguyen T.L. Lien[a], Heikki Kolehmainen[a], Eero Hiltunen[b] and Mousa M. Nazhad[a]

[a]Pulp and Paper Technology, Asian Institute of Technology, Klong Luang, Pathumuthani 12120, Thailand

[b]Helsinki University of Technology, Espoo, FIN-02015, Finland

Introduction

The quality of the papers prepared from chemical and mechanical pulps behave differently by recycling. Recycling deteriorates papermaking quality of chemical pulps (low yield), but it improves papermaking quality of mechanical pulps. The difference by recycling was generally considered as the differences in the chemical composition of the two fiber types. Although this assumption was proposed by different authors [1-4], it has not been examined well. The present work addresses this issue.

Results and Discussion

Fines-free unbleached chemithiremomechanical pulp (CTMP), referred as S0, was selectively delignified with different strength to prepare samples with the removal lignin ratio of 16, 36, 46, 50, 68 and 89% using sodium chlorite and named S16, S36, S46, S50, S68 and S89, respectively. The samples were recycled three times and their tensile strength, density and fiber properties (fiber length and coarseness) were measured and compared with the original CTMP (S0). The chemical composition of pulp fiber cell wall was analyzed using HPLC. The X-ray photoelectron spectroscopy (XPS) and confocal scanning laser microscopy (CSLM) were employed to estimate lignin proportions on the surface of pulp fiber cell wall.

It was observed that the recycling trend of mechanical pulp has completely altered with 16% lignin removal (i.e., 4% residual lignin content) as shown in Figure 1 although xylan content of the samples at this strength of delignification remained intact (see Figures 2 and 3).

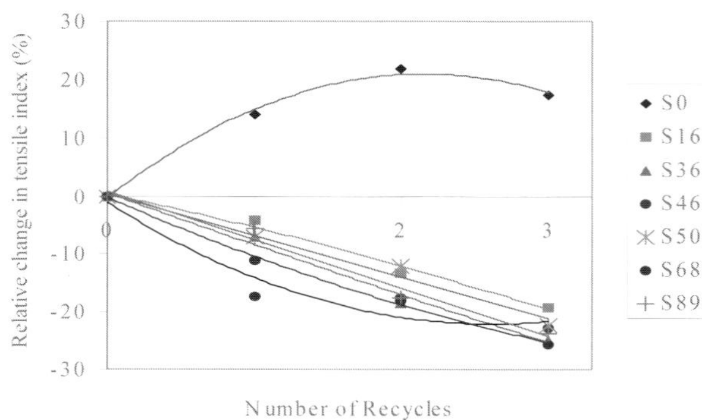

Figure 1 Change in tensile index due to recycling.

Based on this, it was concluded that the difference in the behavior of the two fiber types could not be explained by the xylan content of the fibers.

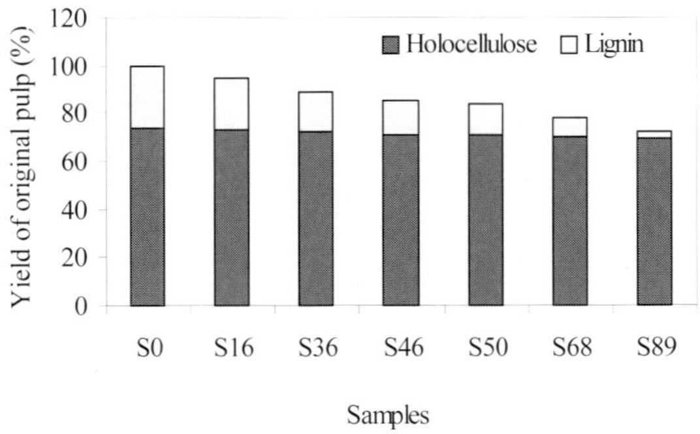

Figure 2 Holocellulose and lignin contents of the samples at different delignification degree.

The lignin content on the surface of fibers of original CTMP (S0) was slightly higher (about 6%) than the fibers of sample S16. The difference in lignin content as detected by XPS is close to the limit of sensitivity; however, the results will be reliable because CSLM supported the data obtained by XPS. The CSLM images showed lower surface lignin

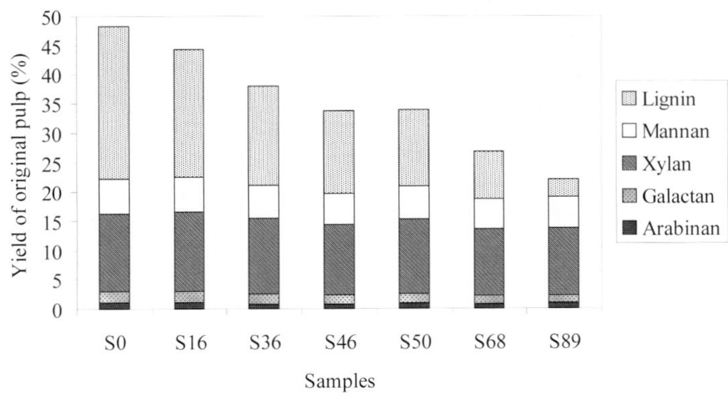

Figure 3 Lignin and hemicellulose contents in samples with various lignin content measured by HPLC.

content for the treated samples compared to a control (Figures 4 and 5). These observations indicate the influence of surface lignin on recycling behavior of the samples. It was previously reported that the surface lignin content was lower after pulping either using kraft or sulphite process [5-7] since the fiber surface delignification occurred at a higher rate than bulk delignification [7].

In the CTMP process, the fibers are separated within the middle lamella zone by synergic action of elevated temperature and mechanical fatigue. The fibers produced by this process are heavily coated with lignin. It is postulated that the surface lignin of fibers is removed at higher rate than the bulk lignin in delignified samples. If this is the case, then the different behavior of the two pulps of S0 and S16 should have some bearings with the loss in surface lignin of the fibers.

Extractive components of the fiber surface of the original CTMP pulp (S0) was almost 50% higher than S16 sample detected by XPS. It has been previously reported that the lignin coated fibers of mechanical pulp are mixed with resin and fatty acids [8]. These materials have been migrated to fiber surface from resin ducts and parenchyma cells in the process of pulping [8]. The presence of extractive components (i.e., resin or fatty acids) on the surface of fibers lowers the surface tension and the wettability causing a reduction in bonding capability of the fibers. These results suggest that the differences in recycling behavior of mechanical pulp compared with chemical pulp could be fairly explained by surface chemical components of fibers (i.e., extractives and lignin).

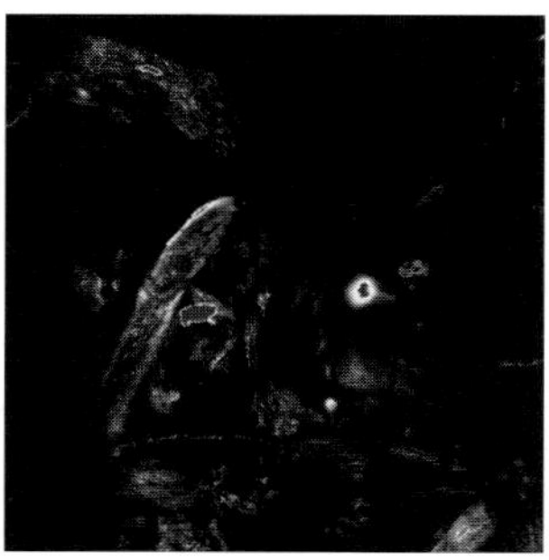

Figure 4 A CSLM image of S0 sample (control).

Figure 5 A CSLM image of S16 sample.

References

[1] Howard, R. C. J. Pulp Paper Sci. 16 (1990) 143.
[2] Nazhad, M. M. and Paszner, L. Tappi J. 77 (1994) 171.
[3] Eastwood, F. G. and Clark, B. "Effects of recycling on paper quality". Trans. BPBIF, Symp. Fiber-water interactions in papermaking, vol. 2 (1978) 835.
[4] Cao, B., Tschirner, U. and Ramaswamy, S. Tappi J. 81 (1998) 119.
[5] Wood, J. and Goring, D. A. I. Pulp Paper Mag. Can. 74 (1973) T309.
[6] Laine, J., Stenius, P., Carlsson, G. and Storm G. Nordic Pulp Paper Res. J. 11 (1996) 201.
[7] Li, K. and Reeve, D. "Surface lignin on chemical wood pulp fibers". Pulp and Paper Center, Chemical Engineering Department, The University of Toronto, Ontario, Canada, Annual report (1999).
[8] Lyne, M. B. and Aspler, J. S., Tappi J. 65 (1982) 98.

Limitation of Fiber Fractionation-Refining Process to Improve Paper Strength Using Recycled OCC Pulp

Mousa M. Nazhad
Pulp and Paper Technology, Asian Institute of Technology, Klong Luang, Pathumuthani 12120, Thailand

Introduction

The linerboard is produced by several plies derived from various pulps. When it is manufactured from pulp of virgin fibers and from pulp containing recycled fibers, it is usually called kraftliner and testliner, respectively. The medium is a single ply product with a fluted (wavy) structure and is sandwiched between two linerboards to form the corrugated board for boxes. The corrugated container has a long history of recycling for many years. The major problem of recycling old corrugated container (OCC) is the decrease of recycled pulp strength. The fractionation of recycled pulp has a potential to upgrade the supplying fibers to enable the production of higher paper grades providing better formulation. One proposed process is to fractionate the recycled OCC pulp into long and short fibers. And the long fibers are refined and recombined with short fibers to be a new stock. However, the results obtained are different among authors as follows [1-4].

Musselmann [1] reported the improvement in strength, freeness and energy consumption by the recombination of refined long fiber fraction and short fiber fraction. However, Nazhad and Awadel-Karim [2] and Sloane and Weinberg [3] reported the effect is limited that the pulp strength from refined feed stock was higher than that from recombined stock (i.e., mixture of refined long fiber fraction and short fiber fraction). A major focus of the work presented here is to assist in settling the on going debate about whether refined feed stock or recombined stock provides better strength properties for OCC pulp. The issue was addressed by comparing strength properties of the sample boards from the two processes.

Results

Overview of recombined stock performance compared with refined feed stock

Figure 1 shows the overall paper quality performance prepared by recombined stock compared with that by the refined feed stock of recycled OCC pulp. In general, the paper strength properties by the recombined stock are lower than those by the refined feed stock (refined unfractionated stock) except for tear strength (Figure 1). Based on the results obtained, the recombined process is not practical and hardly could be recommended for paper or board manufacturers as an option for upgrading OCC pulp.

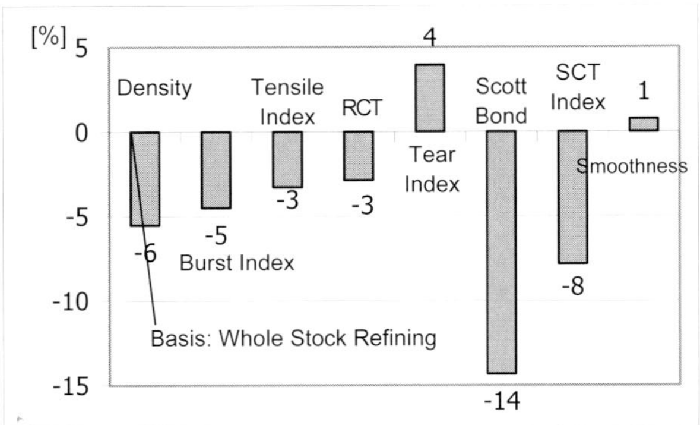

Figure 1 Comparison of paper properties prepared by recombined stock and refined feed stock (refined unfractionated stock).

Comparison of energy consumption between fractionation-refining and full-stream refining processes

One of the advantages in the use of fractionation is the energy saving potential [1, 5]. Net refining energy required for fractionated pulp was 207 kWh/t and that of the feed stock was 256 kWh/t. Therefore, fractionation has brought about 20% energy saving. However, it is inferred that the real energy saving should be less than the value or it could be a complete trade off the energy consumed by a fractionation process. Similar cautious approach in evaluating the net energy consumption was also recommended by earlier workers [3].

Discussion

The major focus of the present study is to compare and contrast the strength properties of the recombined stock (i.e., refined long fiber fraction, then mixed with short fiber fraction) with the refined feed stock. The results obtained, in general, suggested that the refined feed stock yields better quality products (i.e., strength quality). In other words, the recombination process has almost no effect on the paper strength (Figure 1). Nazhad and Awadel-Karim [2] reported similar results using Bauver McNett fractionator and Valley beater in preparation of the recombined stock. The experiments were conducted at the laboratory scale, but using different factionators (laboratory pressure screen and Bauver McNett fractionator) and different refiners (Valley beater and PFI mill). In spite of the differences in the mode of action of the fractionators or the refiners, the results were similar to described above. Our results agreed with the use of a laboratory Packer screen and an Escher Wyss refiner by Sloane and Weinberg [3]. Their observations were also similar to the findings of the present work. Using different ratio of accept to reject in pulp (i.e., 78/22; 63/37; 51/49) to produce the recombined stock did not improve the paper strenght [3]. Therefore, it is plausible to assume that the recombined stock has no effect on the improvement of paper strength. However, the results reported by Musselmann [1] is inconsistent with our results, even if he used a Bauer McNett fractionator and a Jorko refiner at a laboratory scale.

The energy saving, before its correction for the energy required in fractionation, was about 20%. However, the net energy might be less significant after the correction for the energy consumed in fractionation. In addition, the energy saving should be a function of accept to reject ratio in fractionation. The energy saving is expected to be higher with higher accept to reject ratio.

Conclusion

Refined feed stock was stronger in all strength properties than the recombined stock. It was concluded that the fractionation with a scheme to produce pulp superior in quality to the original pulp does not have a promising prospect. However, separate processing has potential to bring about energy saving and to generate pulps with different characteristics.

References

[1] Musselmann, W. "Fractionation of fibrous stock. secondary fiber recycling", Atlanta, USA, TAPPI Press (1993).
[2] Nazhad, M. M. and Awadel-Karim, S. "Limitations of fractionation on upgrading OCC pulp", International Paper Physics Conference Proc., San Diego, USA, (1999) 191.
[3] Sloane, C. M. and Weinberg, G. R. "Product Development & Process Options for Secondary Fiber Liners", 55th Appita Annual Conference, Hobart, Australia (2001) 501.
[4] Kaminski, D. F. "Best practices in fiber fractionation among United States recycled OCC mills", Recycling Symposium, Chicago, USA, TAPPI Press (1997) 483.
[5] Abubakr, S., Scott, G. and Klungness, J. "Fiber fractionation as a method of improving handsheet properties after repeated recycling". Recycling Symposium, Boston, USA, TAPPI Press (1994) 309.

Refining Treatment Could Contribute to Restoring the Strength Properties of OCC Pulp

Mousa M. Nazhad
Pulp and Paper Technology, Asian Institute of Technology, Klong Luang, Pathumuthani 12120, Thailand

Introduction

The volume of recovered paper used globally is almost equal to the volume of new chemical pulp or four times that of mechanical pulp. Recycling has gained a very strong ground indeed. Nowadays, in some countries the paper industry cannot survive without recycled fibers.

Packaging papers and boards have the highest utilization rate in the world among the recycling grade papers and this trend will continue at the same level if the quality requirement for papers is satisfied. Major component of old corrugated container (OCC) pulp is softwood kraft pulp. The loss in tensile or burst strength of chemical pulps is about 30% after a few times recycling [1]. Tear strength, by contrast, has a gain, which is about 30%. These strength changes are caused by the loss in fiber-to-fiber bonds [2].

To overcome the problems of strength decrease in recycled papers, the industry explores some options [2-3]. The pulp refining is one of the options to improve the quality although it can easily be counterproductive if it is not practiced properly [3-4]. Recent developments in refining theory have made it possible to develop desired pulp properties controlling refining parameters [4]. Although the specific edge load (SEL) theory of refining ignores some parameters such as fiber characteristics or detailed pulp flow properties, however it incorporates some critical refining parameters. The theory is widely applied practically for controlling refiners using two parameters: specific energy and intensity [4].

The objective of the present study is to examine the refining potential on improving the strength of recycled papers based on specific edge load theory.

Results and Discussion

Influence of specific energy

Figure 1 shows that the refining can restore some of the strength properties of OCC pulp. Refining the pulp to about 60 kWh/t increases the tensile or burst strength 30% to 40% and reduces tear strength 20% to 30%. DeFoe [5] reported similar alteration on strength properties of OCC pulp by refining. Burst strength of the commercial pulp could be improved about 40% at gross specific energy of 50 kWh/t with a minimum loss in tear strength (about 10%) in a previous report [2]. Lumainen [3] reported a 60% increase in tensile strength at specific energy of 60 kWh/t. The results obtained and the reports suggest that the increase in strength properties of OCC is due to the generation of fines by the refining because the strength is much higher than the virgin strength.

Refining of OCC pulp generates fines and reduces the fiber length (i.e., increase in the population of fibers in one gram of pulp) as in Figure 2. This agrees with previous reports [6, 7]. Increase in fiber population per gram of pulp enhances fiber to fiber bonding because of more surfaces available for bonding by fines [8]. Therefore, it could be concluded that refining could fully restore bonding potential of recycled fibers through generation of fines and fragmentation of fibers, but does not imply the restoration of the original state of the fibers.

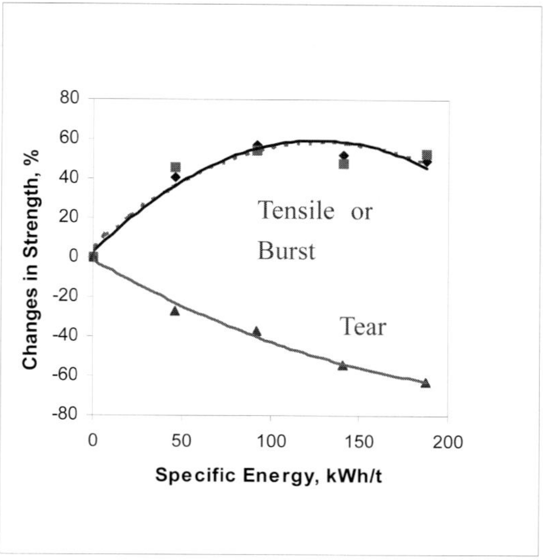

Figure 1 Changes in pulp strength as a function of gross specific energy of refining.

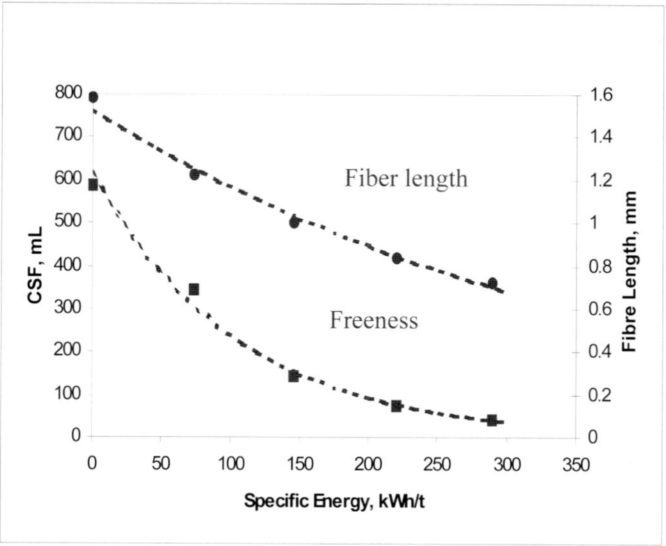

Figure 2 Influence of gross specific energy of refining on freeness and fiber length.

Influence of refining intensity

Figure 3 indicates that the refining intensity does not have influence on tensile strength development. It is assumed that the influence of refining intensity, if there is any, at this range (0 to 100 kWh/t) should be minor, so the effect could not be detected. The literature on the effect of refining intensity on fiber development is abundant, but contradictory.

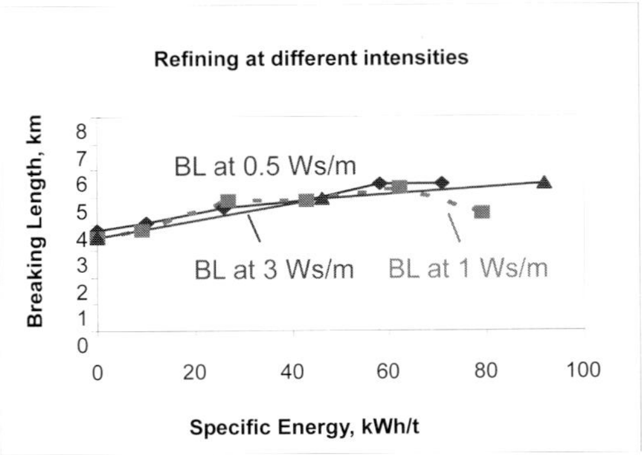

Figure 3. Influence of intensity on tensile index (BL).

Lumainen [9] examined the role of intensity on refining of bleached *Eucalyptus* kraft pulp using three different refiners (laboratory-size disk refiner and two mill-scale conical refiners). As results, the effect of refining intensity was different among refiners as the highest tensile strength obtained at lower intensity by two types of the refiners, and no effect by the other. Two different OCC pulps with the same refiner had two different responses to the given intensity as reported by Defoe [5]. The refining of the present work has been conducted by an Escher Wyss refiner and showed different effects on the pulp [10, 11].

Baker [4] suggested a universal intensity range for refining all types of softwood, hardwood and recycled pulps. Lumainen [8] reported three different optimum refining intensities for bleached *Eucalyptus* kraft pulp.

All of these controversial results share a common wisdom; stressing on a point of view that the refining is more complicated than to be characterized merely with intensity or specific energy.

References

[1] Howard, R.C. and Bichard, W. J. Pulp Paper Sci. 18 (1992) J151.
[2] Nazhad, M.N. and Paszner, L. Tappi J. 77 (1994) 171.
[3] Lumainen, J. Tappi J. 75 (1992) 92.
[4] Baker, C.F., Tappi J. 78 (1995) 147.
[5] DeFoe R.J. Tappi J. 76 (1993) 157.
[6] Szwarcsztajn, E. and Przybysz, K. Cell. Chem. Technol. 10 (1976) 737.
[7] Atwood, D, Paper Technol. Ind. 319 (1990) 2.
[8] Nazhad, M. M., Karchanapoo and Palokangas, A. Appita J. 56 (2003) 61.
[9] Lumainen, J., 'Is the lowest refining intensity the best in low consistency refining of hardwood pulps?', Papermakers Conference, Tappi proceedings (1994) 115.
[10] Seth, R.S., 'Optimizing reinforcement pulps by fracture toughness', Pira 3th Int. Refining Conf., Atlanta, GA, Paper 1 (1995).
[11] Levlin, J-E, 'Some differences in the beating behavior of softwood and hardwood pulps', Intl. Symp. Fundamental Concepts of Refining, Appleton, WI, USA (1980) 51.

Chapter 3

Anatomical Aspects

Formation and Structure of Reaction Wood Fibers Forming No G-Layer in Some Hardwood Species

Nobuo Yoshizawa[a], Futoshi Ishiguri[a], Shinso Yokota[a] and Toshihiro Ona[b]

[a]*Faculty of Agriculture, Utsunomiya University, 350 Mine-machi, Utsunomiya 321-8505, Japan*
[b]*Graduate School of Bioresource and Bioenvironmental Sciences, Kyushu University, Fukuoka 812-8581, Japan*

Introduction

Woody plants develop special tissues, which are called reaction wood, on the inclined stems or branches. Tension wood, reaction wood for hardwoods, can be characterized by the presence of gelatinous (G) -fibers which contain a thickened cellulosic inner layer (G-layer) [1-3]. However, some species of primitive hardwoods have been reported to frequently lack of G-fibers [4-7]. No detailed information is available about the structure and chemical composition of tention wood fibers in these species. In this paper, anatomical features of the tention wood fibers in three species forming no G-layer were observed. In addition, distribution of guaiacyl- and syringyl-type lignins in the cell walls of both normal and tention wood fibers was examined by visible light (VL) combined with microspectrophotometry after the Wiesner and Mäule reactions.

Results and Discussion

Cell wall structure of tention wood fibers formed on branch wood of three hardwood species (*Magnolia obovata*, *Liriodendron tulipifera* and *Paulownia tomentosa*) was examined microspectrophotometrically. In these species, instead of forming G-fibers, low-lignin-content-fibers were formed for xylem on the upper side of blanches in which a radial growth promotion occurred except for *P. tomentosa* and vessels distributed poorly with a small diameter. The secondary wall of wood fibers was consisted of three normal layers, $S_1 + S_2 + S_3$, on the lateral and the lower side of branches, but tension wood fibers on the upper side of branches lacked an S_3 layer. This feature resembles to conifer racheids of compression wood which is reaction wood of softwoods. The S_2 layer of tension wood fibers formed on the upper side of branches exhibited a smaller microfibrillar angle compared to that of the other parts (Table 1).

Microspectrophotometric analysis of the thin cross sections stained by the Mäule and Wiesner color reactions showed a decrease in lignin content of S_2 layer (Figure 1). The wetchemical lignin analysis also indicated a decrease in lignin content with an increase in lignin syringyl / guaiacyl ratio by thioacidolysis (Figure 2), suggesting that syringyl lignin is rich in the S_2 layer of such low-lignin-content-fibers.

Table1 Microfibrillar angle of S2 layer in wood fiber in three hardwood species.

Species	Position	MFA (degree)
M. obovata	RW	15.3
	LW	18.2
	OW	19.7
L. tulipifera	RW	18.2
	LW	20.3
	OW	23.2
P. tomentosa	RW	17.7
	LW	19.3
	OW	19.7

Note: RW, reaction wood; LW, lateral wood; OW, opposite wood.
: MFA, microfibrillar angle.

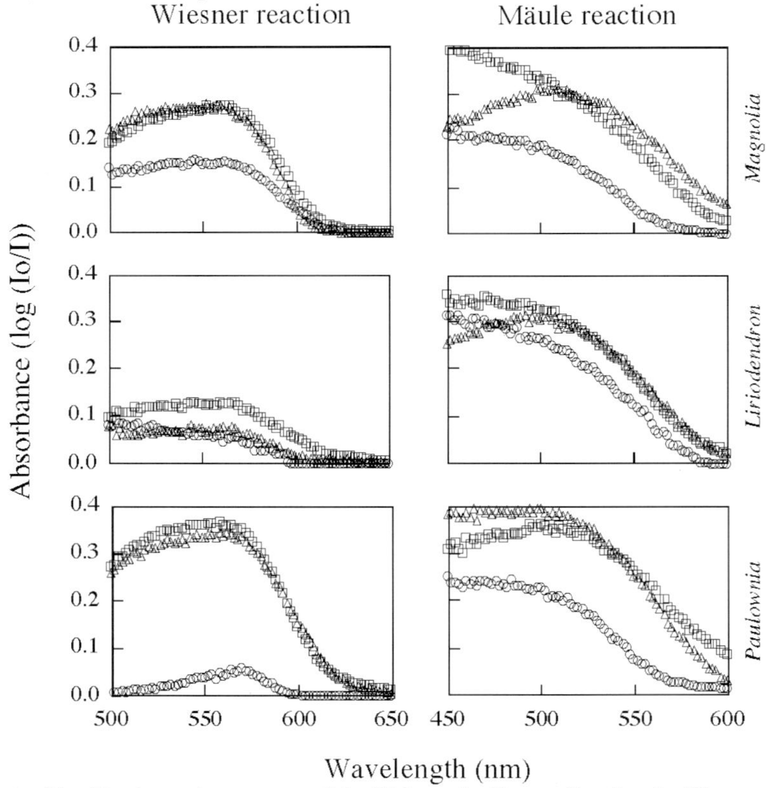

Figure 1 The VL absorption spectra of the S2 layer in fiber walls after the Wiesner and Mäule color reactions. ○; reaction wood, △; lateral wood, □; opposite wood.

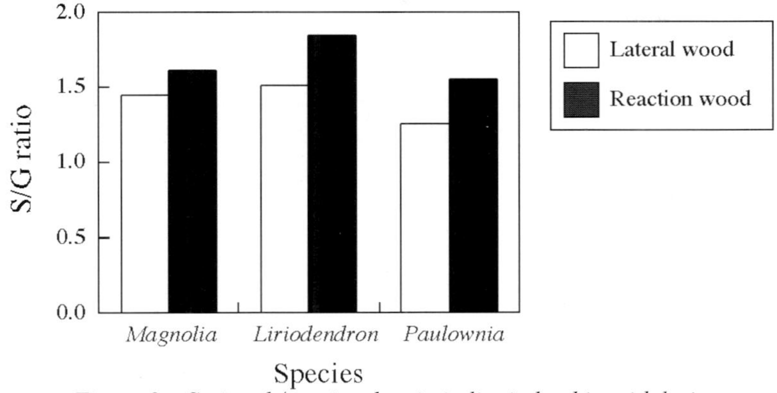

Figure 2 Syringyl / guaiacyl ratio in lignin by thioacidolysis.

References

[1] Onaka, F. Wood Res. (1949) 1.
[2] Dadswell, H. E. and Wardrop, A. B. Holzforschung. 9 (1955) 97.
[3] Côté, W. A. Jr. and Day, A. C. Cellular ultrastructure of woody plants, Syracuse Univ. Press, New York. (1965).
[4] Maylan, B. A. Wood Sci. Technol. 15 (1981) 81.
[5] Timell, T. E. Compression wood in gymnosperms. Vol. I. Springer, Berlin. (1986).
[6] Yoshizawa, N., Satoh, M., Yokota, S. and Idei, T. Wood Sci. Technol. 27 (1993) 1.
[7] Yoshizawa, N., Watanabe, N., Yokota, S. and Idei, T. IAWA J. 14 (1993) 139.

Radial Variation of Cell Morphology in Three *Acacia* Species

Futoshi Ishiguri[a], Shinso Yokota[a], Nobuo Yoshizawa[a] and Toshihiro Ona[b]

[a]*Faculty of Agriculture, Utsunomiya University, 350 Mine-machi, Utsunomiya 321-8505, Japan*
[b]*Graduate School of Bioresource and Bioenvironmental Sciences, Kyushu University, Fukuoka 812-8581, Japan*

Introduction

Acacia species are widely planted for pulp production in south-east Asian countries. It is well known that pulp properties strongly relate to cell morphology of wood [1-3]. In the present study, radial variation of cell morphology (diameter, wall thickness and length in vessel element and wood fiber, and volume of cell types) in two *Acacia* species (*A. mangium* and *A. auriculiformis*) and their hybrid (*A. mangium* x *A. auriculiformis*) was examined.

Materials and Methods

Two trees of each 11-year-old *A. mangium*, *A. auriculiformis* and the hybrid (*A. mangium* x *A. auriculiformis*) were used in this experiment. The height and diameter at breast height of sample trees are shown in Table. The disks of 6 cm in thickness were collected at breast height position. For the cell morphology examination, 2 cm strips were taken through pith to bark from each disk. The cell morphology was determined at every 2 cm from pith to bark. The volume of cell types was determined by point-count technique according to Taylor [4].

Table Summary of sample trees.

Species		Height (m)	D.B.H. (cm)
A. auriculiformis	No. 1	23.6	21.4
	No. 2	23.0	24.8
A. mangium	No. 1	24.6	26.4
	No. 2	22.7	24.4
Hybrid	No. 1	29.6	21.2
	No. 2	29.8	29.3

Results and Discussion

Figure shows volume of cell types in three *Acacia* species. The average volume of vessel element was 16.8, 14.7 and 13.5 % in *A. auriculiformis, A. mangium* and the hybrid, respectively. Almost equal volume of wood fiber was observed between *A. mangium* and the hybrid. However, *A. auriculiformis* showed apparently lower volume of fiber than the other two species. In general, it is thought that wood samples having low volume of vessel element and high volume of wood fiber is suitable for pulp production because more wood fiber is contained in the wood samples. In addition, in *Eucalyptus* woods, Ona et al. [3] reported that decreasing the proportion of ray and axial parenchyma cells within xylem would be an objective in the breeding program to obtain high quality pulp. In the present study, the volumes of parenchyma cells (ray + axial) were 16.8, 12.2 and 14.3 in *A. auriculiformis*, *A. mangium* and the hybrid, respectively. Therefore, *A. mangium* and the hybrid can be suitable for raw materiasl for pulp production.

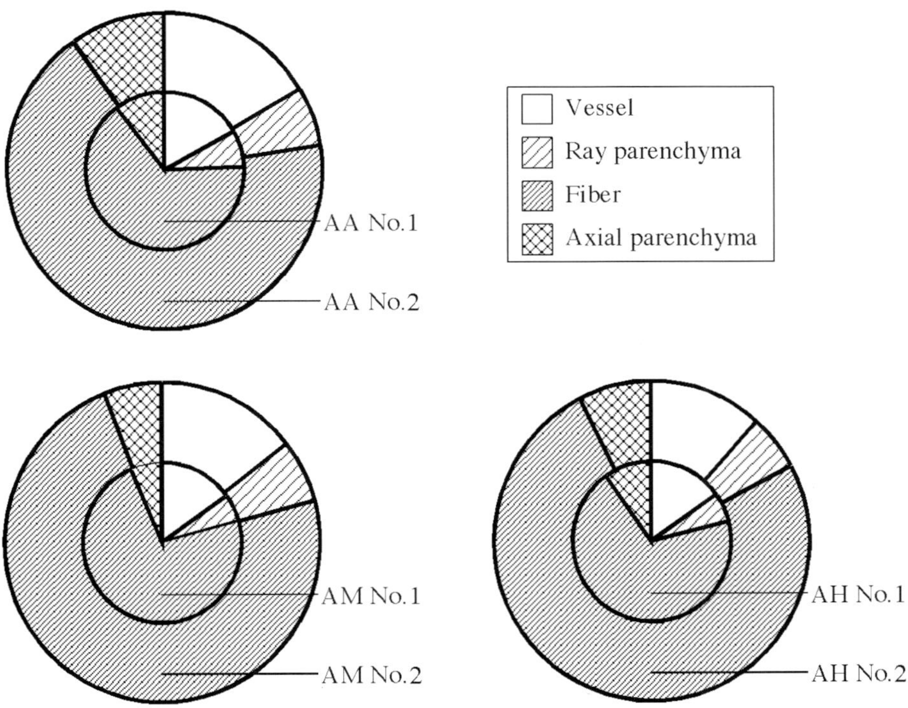

Figure 1 Volume of cell types in three Acacia species. AA, AM and AH indicates A. mangium, A. auriculiformis and the hybrid, respectively.

References

[1] Zobel, B. J. and Buijtenen, J. P. Wood variation, Springer-Verlag, Berlin (1989).
[2] Shimaji, K., Sudo, S. and Harada, H. Mokuzai No Soshiki (Japanese), Morikita Shuppan, Tokyo (1976).
[3] Ona, T., Sonoda, K., Shibata, M., Tamai, Y., Kojima, Y., Ohshima, J., Yokota, S. and Yoshizawa, N. Wood Sci. Technol. 35 (2001) 229.
[4] Taylor, F. W. Appita J. 27 (1973) 171.

Within-Tree Variation of Detailed Fiber Morphology and Its Position Representing the Whole-Tree Value in *Eucalyptus camaldulensis* and *E. globulus*

Jyunichi Ohshima[a], Shinso Yokota[a], Nobuo Yoshizawa[a] and Toshihiro Ona[b]

[a]*Faculty of Agriculture, Utsunomiya University, 350 Mine-machi, Utsunomiya 321-8505, Japan*
[b]*Graduate School of Bioresource and Bioenvironmental Sciences, Kyushu University, Hakozaki, Higashi-ku, Fukuoka 812-8581, Japan*

Introduction

Eucalyptus is one of the major plantation species for pulpwood in many countries. It is of importance to breed trees possessing high pulp yield and quality for the management of pulpwood forests. The selection of elite trees based on their quality has become increasingly significant from both environment issues and the reduction of pulp cost, especially for plantations with short rotations [1]. In the quality breeding program, the increment core method is preferably performed [2]. However, this is an indirect selection method that estimates whole-tree properties using the relationships between wood and pulp properties and requires a core-sampling position that will represent the whole-value. Fiber morphology is an important determinant of pulp properties in *Eucalyptus*. We examined the within-tree variation and the representative height in the trunk to indicate whole-tree values of fiber morphology (length, radial, tangential and average diameters, lumen diameter, and wall thickness) in *Eucalyptus camaldulensis* and *E. globulus* trees [3].

Results and Discussion

Within-tree variations of fiber length in *E. camaldulensis* and *E. globulus* are shown in Figures 1 and 2, respectively. In *E. camaldulensis*, both No. 1 and No. 2 trees had long fibres in the upper and outer parts of the trunk. In particular, the upper parts in No. 1 showed long fibers compared to the other parts. There was a trend of increasing fiber length from pith to bark in the radial direction, and from bottom to top in the axial direction. Tree-to-tree variation in fiber length was large. In contrast to this, the *E. globulus* trees had long fibers in the outer parts of the trunk only. The outer parts at about one-third of the tree height above the ground showed particularly long fibers compared to the other parts. There was a trend of increasing fiber length from pith to bark in the radial direction. A relatively weak trend in the axial direction was observed in the inner parts. On the other hand, tree-to-tree variation was small in *E. globulus*. Consequently, a significant difference between species was observed for fiber length. In the radial direction, the trend of increasing fiber length from pith to bark was reported for *Eucalyptus* [4]. We observed similar trends of increasing fiber length in our samples, which were unaffected by the different growth rates. This suggests the increase in fiber length is based on an increase in the length of the cambial initials with increasing cambial age.

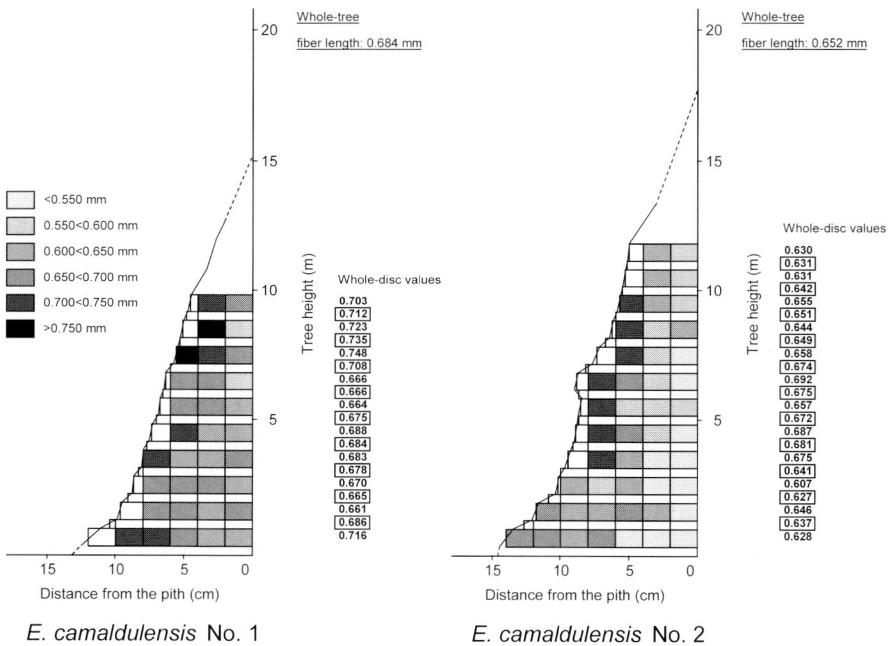

Figure 1 Within-tree variation of fiber length in E. camaldulensis.

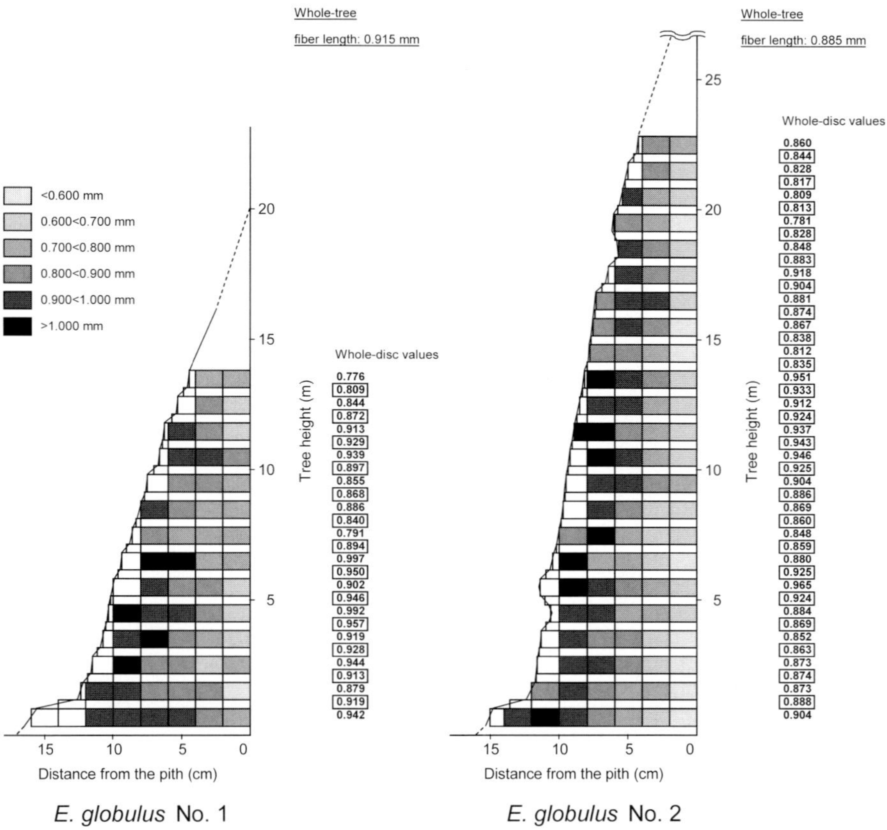

Figure 2 Within-tree variation of fiber length in E. globulus.

Within-tree variation of wall thickness in *E. globulus* is shown in Figure 3. In both species, the trees had high wall thickness in the upper and outer parts of the trunk. The trend of increasing wall thickness was observed from pith to bark in the radial direction, and from bottom to the top in the axial direction in all samples. Significantly, similar trends were observed among samples. Our results are in good agreement with other reports [4] for both species. The bark side and the upper parts of the trunk would have higher pulp yield and better digestibility than other parts. The representative heights used to indicate the whole-tree values of fiber morphology were below 3.3 m above the ground. Representative heights are defined as whole-tree value ± 5% following a previous report [5]. The predicted values of fiber morphology in *E. camaldulensis* and *E. globulus* are summarized in Table. From these results, the most appropriate height to sample an increment core is concluded to be 1.3 and 2.8 m above the ground for *E. camaldulensis* (except fiber wall thickness), and 1.8 m for *E. globulus*. On the other hand, wall thickness can be calculated by the difference of average and lumen diameters. These heights did not relate to the differences in tree height (growth rate) or trends of within-tree fiber morphology variation. The results suggest it is feasible to estimate whole-tree values of fiber morphology, using a single increment core sampled from the representative height. These measurements could be used in *Eucalyptus* tree breeding for the improvement of pulpwood quality.

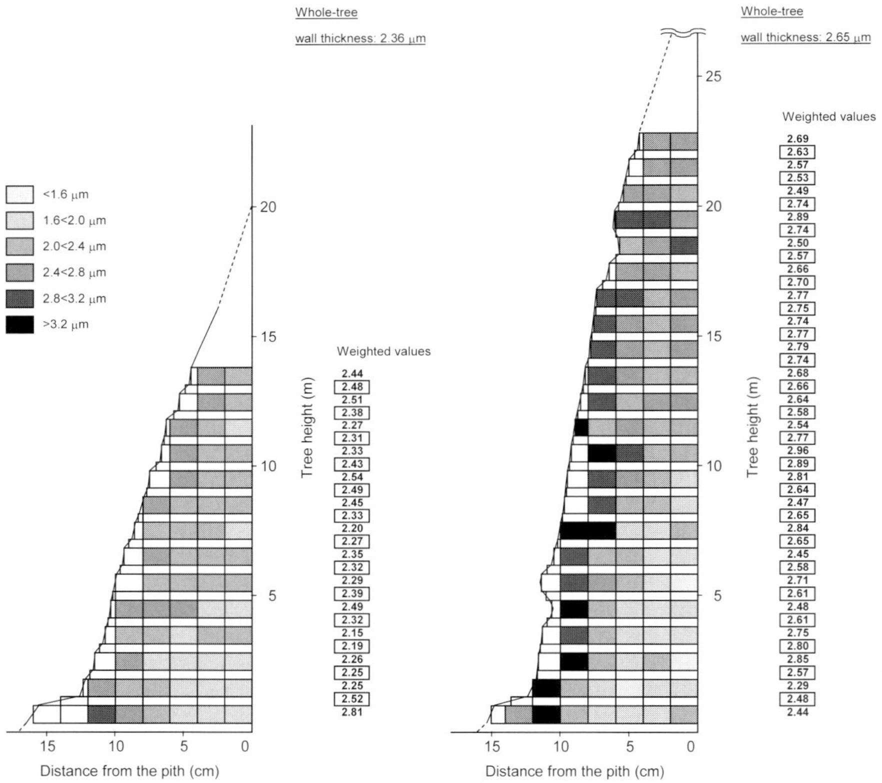

E. globulus No. 1   *E. globulus* No. 2

Figure 3 Within-tree variation of wall thickness in E. globulus.

Table Whole-tree and -disc values of fiber morphology at given heights in E. camaldulensis and E. globulus.

Fiber morphology	Height above ground (m)	*E. camaldulensis*				*E. globulus*			
		No. 1		No. 2		No. 1		No. 2	
		Predicted value (b)	a-b	Predicted value (b)	a-b	Predicted value (b)	a-b	Predicted value (b)	a-b
Length (mm)	Whole-tree value (a)	0.684	–	0.652	–	0.915	–	0.885	–
	3.3	0.684	0.000	0.675	- 0.023	0.919	- 0.004	0.852	0.033
	2.8	0.683	0.001	0.641	0.011	0.928	- 0.013	0.863	0.022
	2.3	0.678	0.006	0.607	0.045	0.944	- 0.029	0.873	0.012
	1.8	0.665	0.019	0.627	0.025	0.913	0.002	0.874	0.011
	1.3	0.661	0.023	0.646	0.006	0.879	0.036	0.873	0.012
	0.8	0.686	- 0.002	0.637	0.015	0.919	- 0.004	0.888	0.003
	0.3	0.716	- 0.032	0.628	0.024	0.942	- 0.027	0.904	- 0.019
Radial diameter (μm)	Whole-tree value (a)	12.7	–	12.6	–	16.4	–	15.6	–
	3.3	12.6	0.1	12.9	- 0.3	17.5	- 1.1	15.9	- 0.3
	2.8	12.7	0.0	13.1	- 0.5	17.0	- 0.6	15.6	0.0
	2.3	12.9	- 0.2	13.4	- 0.8	16.3	0.1	15.3	0.3
	1.8	12.7	0.0	13.3	- 0.7	16.6	- 0.2	15.5	0.1
	1.3	12.6	0.1	13.2	- 0.6	17.0	- 0.6	15.7	- 0.1
	0.8	13.2	- 0.5	13.4	- 0.8	16.6	- 0.2	16.6	- 1.0
	0.3	13.9	- 1.2	13.8	- 1.2	15.8	0.6	17.6	- 2.0
Tangential diameter (μm)	Whole-tree value (a)	11.3	–	12.0	–	13.5	–	13.6	–
	3.3	10.9	0.4	11.6	0.4	14.3	- 0.8	14.2	- 0.6
	2.8	11.1	0.2	12.0	0.0	14.1	- 0.6	14.0	- 0.4
	2.3	11.3	0.0	12.4	- 0.4	13.8	- 0.3	13.7	- 0.1
	1.8	11.4	- 0.1	12.0	0.0	13.6	- 0.1	13.7	- 0.1
	1.3	11.5	- 0.2	11.6	0.4	13.5	0.0	13.7	- 0.1
	0.8	11.8	- 0.5	12.0	0.0	13.5	0.0	13.9	- 0.3
	0.3	12.0	- 0.7	12.2	- 0.2	13.1	0.4	13.7	- 0.1
Average diameter (μm)	Whole-tree value (a)	12.0	–	12.3	–	14.9	–	14.6	–
	3.3	11.7	0.3	12.3	0.0	15.9	- 1.0	15.1	- 0.5
	2.8	11.9	0.1	12.6	- 0.3	15.6	- 0.7	14.8	- 0.2
	2.3	12.1	- 0.1	12.9	- 0.6	15.1	- 0.2	14.5	0.1
	1.8	12.1	- 0.1	12.6	- 0.3	15.1	- 0.2	14.6	0.0
	1.3	12.1	- 0.1	12.4	- 0.1	15.2	- 0.3	14.7	- 0.1
	0.8	12.5	- 0.5	12.7	- 0.4	15.1	- 0.2	15.3	- 0.7
	0.3	12.9	- 0.9	13.0	- 0.7	14.5	0.4	15.6	- 1.0
Wall thickness (μm)	Whole-tree value (a)	1.88	–	1.84	–	2.36	–	2.65	–
	3.3	2.05	- 0.17	1.79	0.05	2.15	0.21	2.75	- 0.10
	2.8	1.89	- 0.01	1.74	0.10	2.19	0.17	2.80	- 0.15
	2.3	1.72	0.16	1.70	0.14	2.26	0.10	2.85	- 0.20
	1.8	1.75	0.13	1.71	0.13	2.25	0.11	2.57	0.08
	1.3	1.76	0.12	1.71	0.13	2.25	0.11	2.29	0.36
	0.8	1.79	0.09	1.78	0.06	2.52	- 0.16	2.48	0.17
	0.3	1.84	0.04	1.90	- 0.06	2.81	- 0.45	2.44	0.21
Lumen diameter (μm)	Whole-tree value (a)	7.98	–	8.37	–	9.85	–	8.90	–
	3.3	7.34	0.64	8.43	- 0.06	11.26	- 1.41	9.12	- 0.22
	2.8	7.87	0.11	8.82	- 0.45	10.81	- 0.96	8.74	0.16
	2.3	8.44	- 0.46	9.21	- 0.84	10.20	- 0.35	8.37	0.53
	1.8	8.33	- 0.35	8.97	- 0.60	10.23	- 0.38	9.09	- 0.19
	1.3	8.29	- 0.31	8.70	- 0.33	10.35	- 0.50	9.80	- 0.90
	0.8	8.69	- 0.71	8.91	- 0.54	9.60	0.25	9.91	- 1.01
	0.3	8.99	- 1.01	8.96	- 0.59	8.41	1.44	10.35	- 1.45

References

[1] Ona, T. and Sonoda, T. Cell Commun. 2 (1996) 3.
[2] Downes, G., Hudson, I., Raymond, C., Dean, G., Michell, T., Schimleck, L., Evans, R., and Muneri, A. Sampling plantation eucalyptus for wood and fibre poperties, CSIRO Publications, Melbourne, Australia (1997).
[3] Ohshima, J., Yokota, S., Yoshizawa, N. and Ona, T. Appita J. (2003) *in press*.
[4] Wilks, J. IAWA J. 9 (1988) 13.
[5] Ona, T., Sonoda, T., Ito, K., Shibata, M., Tamai, Y. and Kojima, Y. Appita J. 49 (1996) 325.

Within-Tree Variation of Vessel Morphology and Frequency and Representative Heights for Estimating the Whole-Tree Values in *Eucalyptus camaldulensis* and *E. globulus*

Jyunichi Ohshima[a], Kodai Adachi[a], Shinso Yokota[a], Nobuo Yoshizawa[a] and Toshihiro Ona[b]

[a]*Faculty of Agriculture, Utsunomiya University, 350 Mine-machi, Utsunomiya 321-8505, Japan*
[b]*Graduate School of Bioresource and Bioenvironmental Sciences, Kyushu University, Hakozaki, Higashi-ku, Fukuoka 812-8581, Japan*

Introduction

Often vessels have a major effect on causing the paper properties of *Eucalyptus* [1-3]. They are picked from the paper during printing, causing a non-uniform paper surface, which affects the distribution of ink [1]. Vessel morphology and frequency are important selection indices in the quality breeding program for pulp production. However, the most appropriate core-sampling position has not as yet been reported for vessel morphology and frequency on a whole-tree basis. We examined and the representative height in the trunk to indicate whole-tree values of vessel morphology (length, radial, tangential and average diameters, and wall thickness) and frequency through the within-tree variation in *Eucalyptus camaldulensis* and *E. globulus* trees [4].

Results and Discussion

Within-tree variations of tangential diameter in *E. camaldulensis* and *E. globulus* are shown in Figures 1 and 2, respectively. The trends of within-tree variations of radial and average diameters in both species are very similar to that of tangential diameter. Both *E. camaldulensis* trees had high values for all diameters in the upper and outer parts of the trunk. There were trends of increasing diameters from pith to bark in the radial direction, and from bottom to top in the axial direction. Compared to this, the *E. globulus* trees had high values for all diameters in the outer parts of the trunk only. There were trends of increasing diameters from pith to bark in the radial direction, but relatively uniform trends of increasing diameters from bottom to top in the axial direction were observed only in the inner parts of the trunk. No clear trends were observed in outer parts of the trunk in the axial direction. Consequently, significant between species differences were observed in all diameters in the axial direction. In the radial direction, the trend of increasing vessel tangential diameter from pith to bark was reported for *Eucalyptus* [5]. The trend of increasing vessel diameter in the radial direction may be characteristic for juvenile wood of *Eucalyptus*. The outer parts of the trunk in *E. globulus* are expected to have higher pulp yield and better digestibility than other parts because tangential and average diameters were significantly related to pulp yield (positively) and to digestibility (negatively) [3].

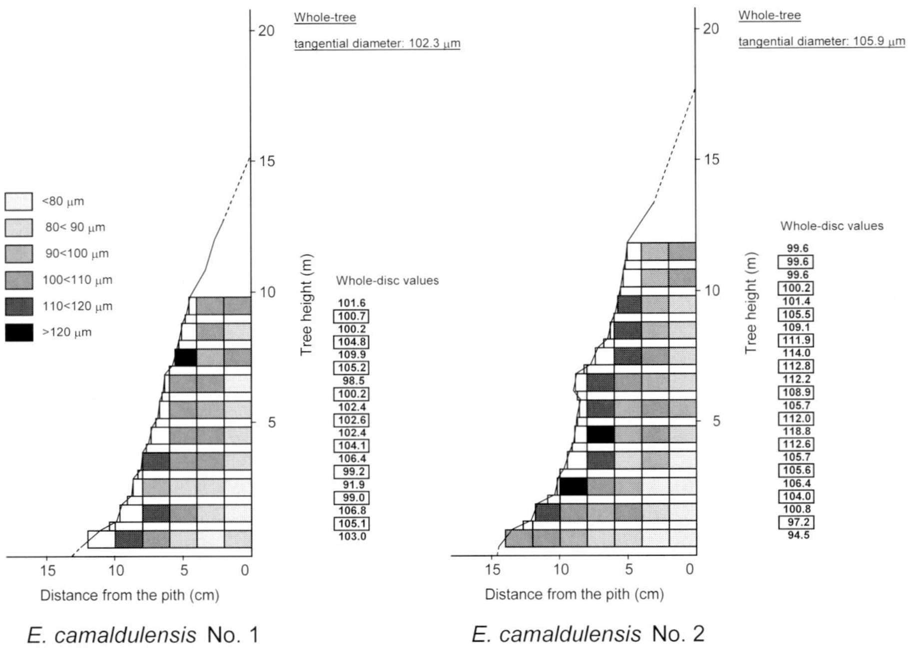

Figure 1   Within-tree variation of tangential diameter in E. camaldulensis.

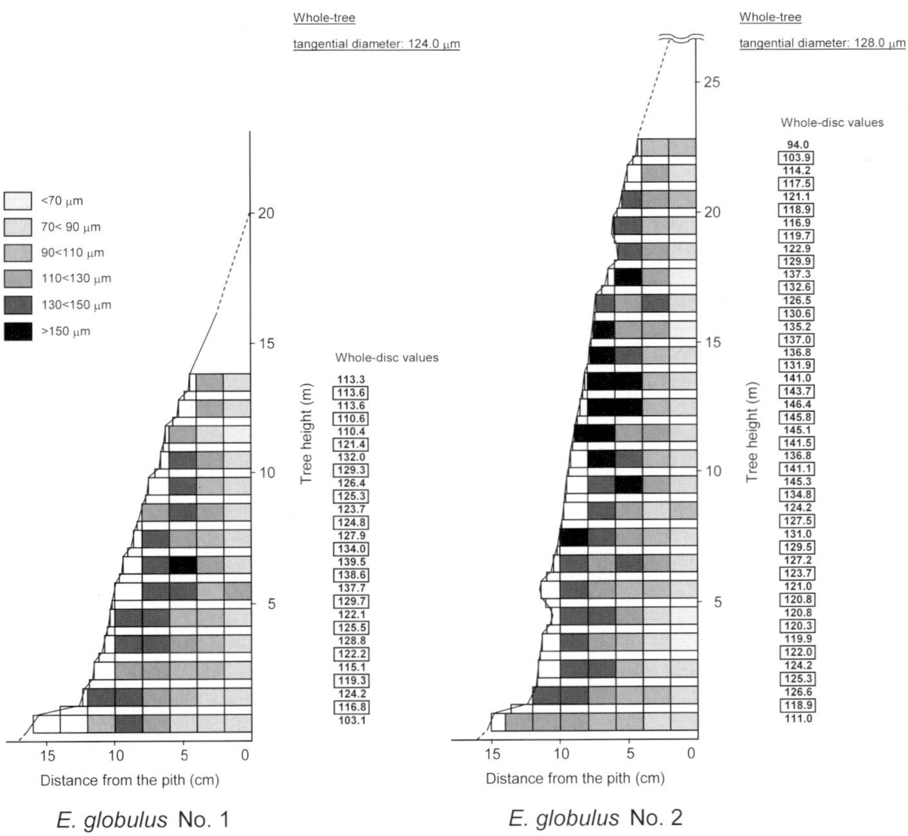

Figure 2 Within-tree variation of tangential diameter in E. globulus.

Within-tree variation of vessel frequency in *E. globulus* is shown in Figure 3. In both species, the trees had high frequency in the inner parts of the trunk. In the radial direction, the trend of decreasing frequency was observed from pith to bark. No clear trend was observed in the axial direction. The trend of decreasing frequency from pith to bark in the radial direction was reported in *Eucalyptus* [5] similar to our results. The outer parts of the trunk would have higher printability because of reduced vessel picking.

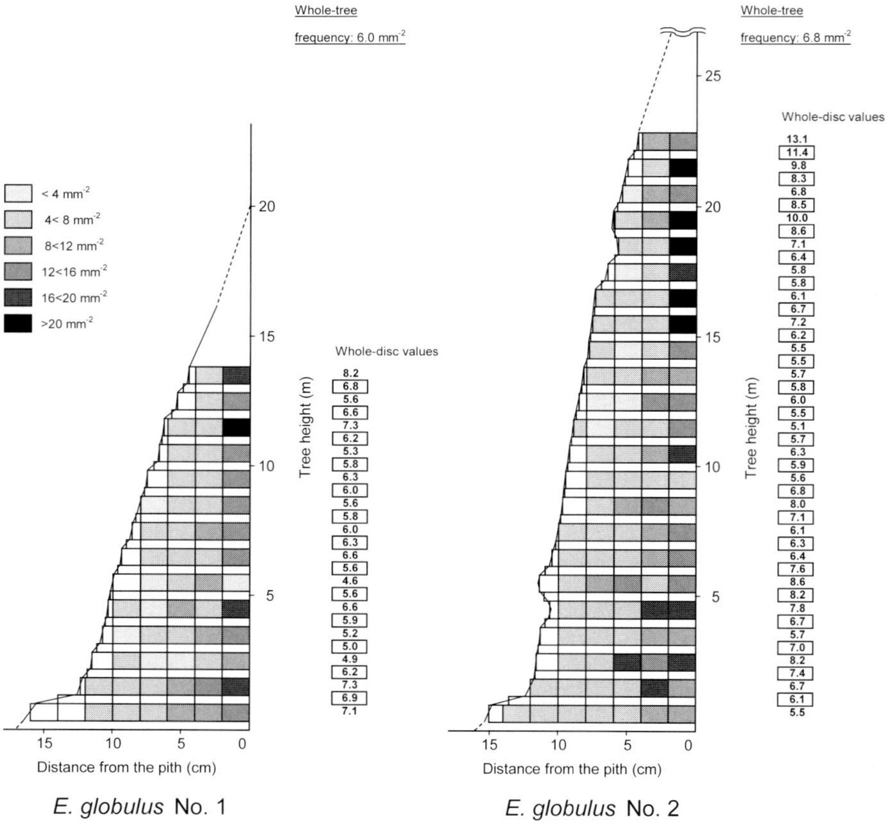

Figure 3 Within-tree variation of frequency in E. globulus.

The representative heights used to indicate the whole-tree values of fiber morphology were below 3.3 m above the ground. Representative heights are defined as whole-tree value ± 5% following a previous report [6]. The predicted values of vessel morphology and frequency in *E. camaldulensis* and *E. globulus* are summarized in Table. From these results, the most appropriate height to sample an increment core is concluded to be 2.8 m above the ground for *E. camaldulensis* (except vessel wall thickness and frequency), and 1.8 m for *E. globulus* (except frequency). No representative height for frequency was obtained for both species, probably because of the large within-tree variation of frequency in the radial direction compared to the other traits. These results suggest it is possible to estimate whole-tree values of vessel morphology using a single increment core sampled at the most appropriate height. It is expected that these measurements will be used in *Eucalyptus* tree breeding for the improvement of pulpwood quality.

Table Whole-tree and -disc values of vessel morphology at given heights in E. camaldulensis and E. globulus.

Vessel morphology	Height above ground (m)	*E. camaldulensis*				*E. globulus*			
		No. 1		No. 2		No. 1		No. 2	
		Predicted value (b)	a-b	Predicted value (b)	a-b	Predicted value (b)	a-b	Predicted value (b)	a-b
Length (mm)	Whole-tree value (a)	0.221	–	0.202	–	0.254	–	0.279	–
	3.3	0.240	- 0.019	0.185	0.017	0.269	- 0.02	0.250	- 0.029
	2.8	0.223	- 0.002	0.197	0.005	0.273	- 0.02	0.277	- 0.002
	2.3	0.207	0.014	0.213	- 0.011	0.277	- 0.02	0.303	- 0.024
	1.8	0.205	0.016	0.203	0.001	0.265	- 0.01	0.275	0.004
	1.3	0.200	0.021	0.189	0.013	0.251	0	0.246	0.033
	0.8	0.200	0.021	0.185	0.017	0.249	0.01	0.231	0.048
	0.3	0.203	0.018	0.177	0.025	0.252	0	0.219	0.060
Radial diameter (μm)	Whole-tree value (a)	138.1	–	150.9	–	184.3	–	177.3	–
	3.3	141.0	- 2.9	153.9	- 3.0	198.2	- 13.9	177.9	- 0.6
	2.8	131.7	6.4	155.7	- 4.8	186.5	- 2.2	174.4	2.9
	2.3	122.3	15.8	159.0	- 8.1	173.2	11.1	171.1	6.2
	1.8	134.2	3.9	153.8	- 2.9	180.3	4.0	174.7	2.6
	1.3	147.3	- 9.2	147.7	3.2	188.6	- 4.3	178.4	- 1.1
	0.8	145.3	- 7.2	143.9	7.0	174.6	9.7	172.2	5.1
	0.3	142.4	- 4.3	139.9	11.0	149.7	34.6	169.6	7.7
Tangential diameter (μm)	Whole-tree value (a)	102.3	–	105.9	–	124.0	–	128.0	–
	3.3	106.4	- 4.1	105.7	0.2	128.8	- 4.8	119.9	8.1
	2.8	99.2	3.1	105.6	0.3	122.2	1.8	122.0	6.0
	2.3	91.9	10.4	106.4	- 0.5	115.1	8.9	124.2	3.8
	1.8	99.0	3.3	104.0	1.9	119.3	4.7	125.3	2.7
	1.3	106.8	- 4.5	100.8	5.1	124.2	- 0.2	126.6	1.4
	0.8	105.1	- 2.8	97.2	8.7	116.8	7.2	118.9	9.1
	0.3	103.0	- 0.7	94.5	11.4	103.1	20.9	111.0	17.0
Average diameter (μm)	Whole-tree value (a)	120.2	–	128.1	–	154.1	–	152.6	–
	3.3	123.7	- 3.5	129.8	- 1.7	163.5	- 9.4	146.0	6.6
	2.8	115.4	4.8	130.7	- 2.6	154.3	- 0.2	146.7	5.9
	2.3	107.1	13.1	132.7	- 4.6	144.1	10.0	147.6	5.0
	1.8	116.6	3.6	128.9	- 0.8	149.8	4.3	150.0	2.6
	1.3	127.1	- 6.9	124.2	3.9	156.4	- 2.3	152.5	0.1
	0.8	125.2	- 5.0	120.6	7.5	145.7	8.4	145.5	7.1
	0.3	122.7	- 2.5	117.2	10.9	126.4	27.7	140.3	12.3
Wall thickness (μm)	Whole-tree value (a)	3.8	–	3.5	–	3.3	–	3.2	–
	3.3	4.0	- 0.2	3.2	0.3	3.3	0.0	3.2	0.0
	2.8	3.9	- 0.1	3.3	0.2	3.2	0.1	3.2	0.0
	2.3	3.8	0.0	3.3	0.2	3.2	0.1	3.2	0.0
	1.8	3.8	0.0	3.3	0.2	3.2	0.1	3.2	0.0
	1.3	3.9	- 0.1	3.4	0.1	3.3	0.0	3.2	0.0
	0.8	3.8	0.0	3.3	0.2	3.4	- 0.1	3.2	0.0
	0.3	3.7	0.1	3.3	0.2	3.4	- 0.1	3.1	0.1
Frequency (No./mm^2)	Whole-tree value (a)	14.4	–	10.4	–	6.0	–	6.8	–
	3.3	14.4	0.0	9.4	1.0	5.2	0.8	5.7	1.1
	2.8	14.9	- 0.5	9.4	1.0	5.0	1.0	7.0	- 0.2
	2.3	15.5	- 1.1	9.3	1.1	4.9	1.1	8.2	- 1.4
	1.8	15.1	- 0.7	9.4	1.0	6.2	- 0.2	7.4	- 0.6
	1.3	14.6	- 0.2	9.6	0.8	7.3	- 1.3	6.7	0.1
	0.8	13.2	1.2	9.8	0.6	6.9	0.9	6.1	0.7
	0.3	11.5	2.9	10.5	- 0.1	7.1	- 1.1	5.5	1.3

References

[1] Colley, J. Paper Technology 14 (1973) 293.
[2] Ogata, Y. Jpn. Tappi 32 (1978) 377.
[3] Ona, T., Sonoda, T., Ito, K., Shibata, Y., M., Tamai, Y., Kojima, Y., Ohshima, J., Yokota, S. and Yoshizawa, N. Wood Sci. Technol. 35 (2001) 229.
[4] Ohshima, J., Yokota, S., Yoshizawa, N. and Ona, T. Appita J. (2003) *in press*.
[5] Wilks, J. IAWA J. 9 (1988) 13.
[6] Ona, T., Sonoda, T., Ito, K., Shibata, M., Tamai, Y. and Kojima, Y. Appita J. 49 (1996) 325.

Representative Heights Assessing Whole-Tree Values and the Within-Tree Variations of Derived Wood Properties in *Eucalyptus camaldulensis* and *E. globulus*

Jyunichi Ohshima[a], Shinso Yokota[a], Nobuo Yoshizawa[a] and Toshihiro Ona[b]

[a]*Faculty of Agriculture, Utsunomiya University, 350 Mine-machi, Utsunomiya 321-8505, Japan*
[b]*Graduate School of Bioresource and Bioenvironmental Sciences, Kyushu University, Hakozaki, Higashi-ku, Fukuoka 812-8581, Japan*

Introduction

The quality breeding for high pulp yield and strength in the pulp production has attracted interest especially for *Eucalyptus*. The increment core method without cutting a tree is effective and efficient in the practical quality breeding program [1], but this is an indirect selection method that requires the prediction of whole-tree properties using clear implications between wood and pulp properties. Derived wood properties, such as Runkel ratio, solid factor, vessel R/T ratio and fiber coarseness, have been recognized an important traits for pulp and paper properties in *Eucalyptus* [2]. We examined the within-tree variations of derived wood properties in *E. camaldulensis* and *E. globulus* trees with the aim of determining the representative height in the trunk to indicate whole-tree values.

Results and Discussion

Within-tree variation of solids factor in *E. camaldulensis* is shown in Figure 1. In *E. camaldulensis*, both No. 1 and No. 2 trees had high solids factor in the upper and outer parts of the trunk although the inner parts of tree No.2 showed lower values than No.1. A trend of increasing solids factor was observed from pith to bark in the radial direction, and from bottom to top in the axial direction. In contrast, both *E. globulus* trees had high solids factor in the outer parts of the trunk only. A trend of increasing solids factor was observed from pith to bark in the radial direction, however, relatively uniform trend of increasing solids factor from bottom to top was observed in the axial direction. Consequently, significant difference between species was observed for solids factor. From high negative relationship between solids factor and breaking length [2], the upper and outer parts of the trunk in *E. camaldulensis* would have lower breaking length than other parts.

Within-tree variations of fiber coarseness in *E. camaldulensis* and *E. globulus* are shown in Figures 2 and 3, respectively. Both *E. camaldulensis* trees had high fiber coarseness in the upper and outer parts of the trunk. A trend of increasing fiber coarseness was observed from pith to bark in the radial direction, except the upper parts of tree No. 2. However, no clear trend was observed in the axial direction. Compared to this, the *E. globulus* trees had high fiber coarseness in the outer parts of the trunk only. In particular, the outer parts of tree No. 2 showed high values compared to the other parts of the trunk. A trend of increasing fiber coarseness was observed from pith to bark in the radial direction, but relatively uniform trend of increasing fiber coarseness from bottom to top in the axial direction was observed. Consequently, fiber coarseness was significantly different between species. In the radial direction, the trend of increasing fiber coarseness from pith to bark was reported in *E. globulus* [3]. We observed similar trends of increasing fiber coarseness in this study. *E. camaldulensis* would have superior paper conformability to *E. globulus* because of lower fiber coarseness (whole-tree fiber coarseness: 0.092, 0.094 mg/m in *E. camaldulensis*; 0.145, 0.154 mg/m in *E. globulus*).

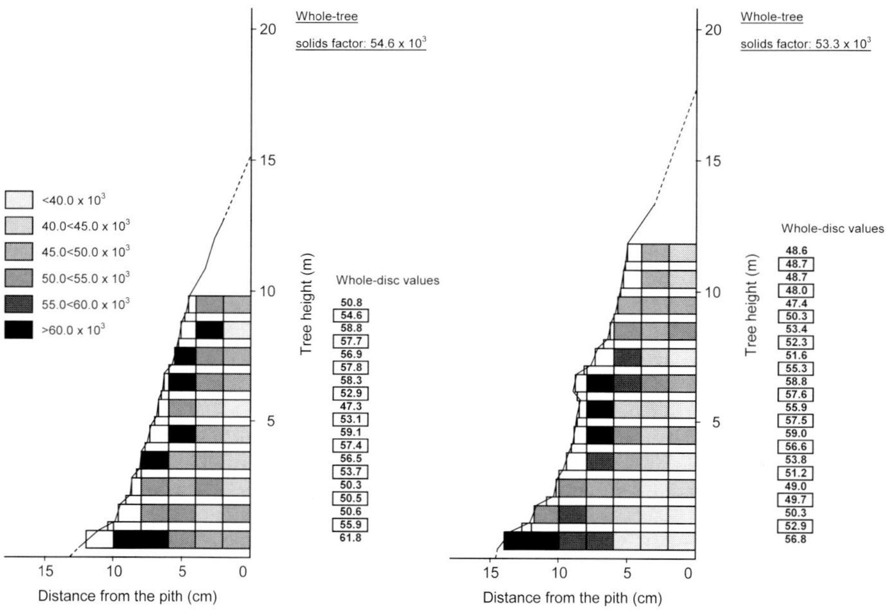

E. camaldulensis No. 1 *E. camaldulensis* No. 2

Figure 1 Within-tree variation of solids factor in E. camaldulensis.

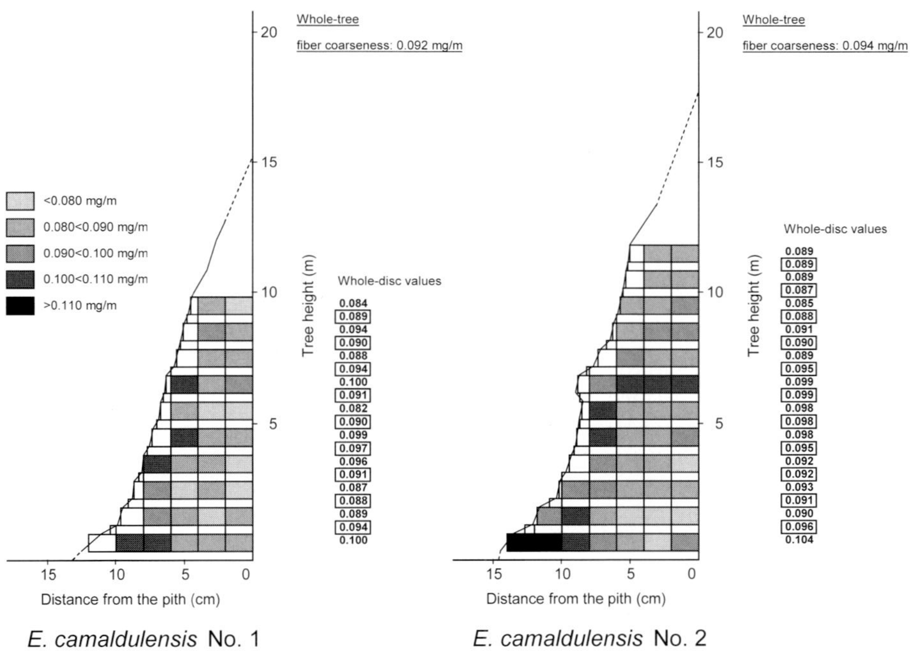

Figure 2 Within-tree variation of fiber coarseness in E. camaldulensis.

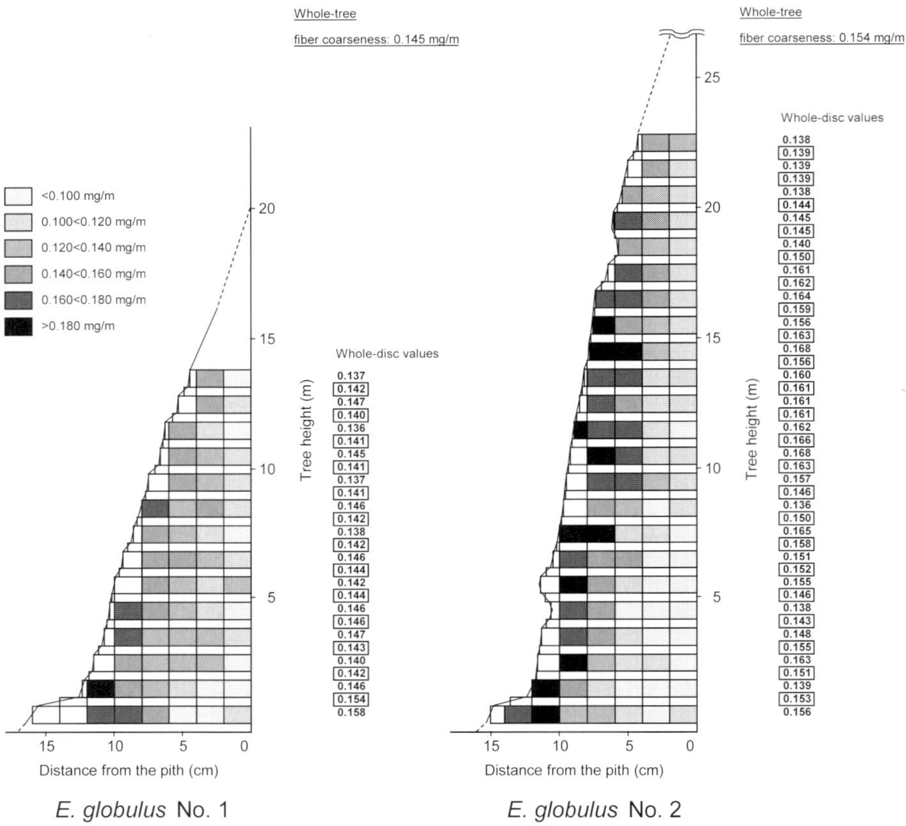

Figure 3 Within-tree variation of fiber coarseness in E. globulus.

The representative heights used to indicate the whole-tree values of derived wood properties were below 3.3 m above the ground. Representative heights are defined as whole-tree value ± 5% following a previous report [4]. The predicted values of derived wood properties in *E. camaldulensis* and *E. globulus* are summarized in Table. From these results, the most appropriate height to sample an increment core is concluded to be 2.8 m above the ground for *E. camaldulensis* (except Runkel ratio, Luce's shape factor, and wall coverage ratio), and 1.8 m for *E. globulus* (except Runkel ratio and wall coverage ratio). No representative heights for Runkel ratio and wall coverage ratio were obtained for both species, probably because of the large difference between whole-tree and –disc values resulted by relatively small or unclear within-tree variation. These results suggest that using a single increment core, sampled at the most appropriate height, it is feasible to estimate whole-tree values of derived wood properties. These measurements and determinations could be used in *Eucalyptus* tree breeding to improve the pulpwood quality.

Table Whole-tree and -disc values of derived wood properties at given heights in E. camaldulensis and E. globulus.

Traits	Height above ground (m)	*E. camaldulensis*				*E. globulus*			
		No. 1		No. 2		No. 1		No. 2	
		Predicted value (b)	a-b	Predicted value (b)	a-b	Predicted value (b)	a-b	Predicted value (b)	a-b
Runkel ratio	Whole-tree value (a)	0.54	–	0.50	–	0.56	–	0.74	–
(-)	3.3	0.66	- 0.12	0.47	0.03	0.43	0.13	0.78	- 0.04
	2.8	0.56	- 0.02	0.44	0.06	0.47	0.09	0.81	- 0.07
	2.3	0.46	0.08	0.41	0.09	0.52	0.04	0.84	- 0.10
	1.8	0.47	0.07	0.43	0.07	0.51	0.05	0.71	0.03
	1.3	0.48	0.06	0.44	0.06	0.50	0.06	0.59	0.15
	0.8	0.46	0.08	0.44	0.06	0.64	- 0.08	0.64	0.10
	0.3	0.45	0.09	0.48	0.02	0.82	- 0.26	0.58	0.16
Flexbility coefficient	Whole-tree value (a)	0.66	–	0.68	–	0.65	–	0.61	–
(-)	3.3	0.62	0.04	0.68	0.00	0.70	- 0.05	0.60	0.01
	2.8	0.66	0.00	0.70	- 0.02	0.69	- 0.04	0.59	0.02
	2.3	0.69	- 0.03	0.71	- 0.03	0.67	- 0.02	0.58	0.03
	1.8	0.69	- 0.03	0.71	- 0.03	0.67	- 0.02	0.62	- 0.01
	1.3	0.69	- 0.03	0.70	- 0.02	0.68	- 0.03	0.66	- 0.05
	0.8	0.69	- 0.03	0.70	- 0.02	0.63	0.02	0.65	- 0.04
	0.3	0.69	- 0.03	0.69	- 0.01	0.58	0.07	0.66	- 0.05
Wall coverage ratio	Whole-tree value (a)	0.34	–	0.32	–	0.35	–	0.39	–
(-)	3.3	0.38	- 0.04	0.32	0.00	0.30	0.05	0.40	- 0.01
	2.8	0.34	0.00	0.30	0.02	0.31	0.04	0.41	- 0.02
	2.3	0.31	0.03	0.29	0.03	0.33	0.02	0.42	- 0.03
	1.8	0.31	0.03	0.29	0.03	0.33	0.02	0.38	0.01
	1.3	0.31	0.03	0.30	0.02	0.32	0.03	0.34	0.05
	0.8	0.31	0.03	0.30	0.02	0.37	- 0.02	0.35	0.04
	0.3	0.31	0.03	0.31	0.01	0.42	- 0.07	0.34	0.05
Luce's shape factor	Whole-tree value (a)	0.39	–	0.37	–	0.40	–	0.46	–
(-)	3.3	0.44	- 0.05	0.36	0.01	0.34	0.06	0.46	- 0.00
	2.8	0.40	- 0.01	0.34	0.03	0.36	0.04	0.48	- 0.02
	2.3	0.35	0.04	0.33	0.04	0.38	0.02	0.50	- 0.04
	1.8	0.36	0.03	0.34	0.03	0.38	0.02	0.44	0.02
	1.3	0.36	0.03	0.34	0.03	0.37	0.03	0.39	0.07
	0.8	0.35	0.04	0.35	0.02	0.43	- 0.03	0.41	0.05
	0.3	0.35	0.04	0.36	0.01	0.50	- 0.10	0.40	0.06
Slenderness ratio	Whole-tree value (a)	57.2	–	53.0	–	61.4	–	60.6	–
(-)	3.3	58.6	- 1.4	55.0	- 2.0	57.9	3.5	57.0	3.6
	2.8	57.1	0.1	51.1	1.9	59.9	1.5	58.6	2.0
	2.3	55.3	1.9	47.2	5.8	62.9	1.5	60.1	0.5
	1.8	55.2	2.0	49.7	3.3	60.7	0.7	59.7	0.9
	1.3	54.9	2.3	52.2	0.8	58.0	3.4	59.2	1.4
	0.8	54.9	2.3	50.1	2.9	61.4	0.0	58.1	2.5
	0.3	55.4	1.8	48.2	4.8	65.3	- 3.9	57.7	2.9
Solids factor	Whole-tree value (a)	54.6	–	53.3	–	116.1	–	121.5	–
(-)(x 10^{-3})	3.3	56.5	- 1.9	53.8	- 0.5	117.0	- 0.9	124.9	- 3.4
	2.8	53.7	0.9	51.2	2.1	116.4	- 0.3	125.3	- 3.8
	2.3	50.3	4.3	49.0	4.3	116.5	- 0.4	125.7	- 4.2
	1.8	50.5	4.1	49.7	3.6	114.3	1.8	116.7	4.8
	1.3	50.9	3.7	50.3	3.0	112.7	3.4	107.7	13.8
	0.8	55.9	- 1.3	52.9	0.4	123.9	- 7.8	121.5	0.0
	0.3	61.8	- 7.2	56.8	- 3.5	129.4	- 13.3	125.9	- 4.4
Vessel diameter R/T ratio	Whole-tree value (a)	1.35	–	1.44	–	1.49	–	1.39	–
(-)	3.3	1.33	0.02	1.50	- 0.06	1.54	- 0.05	1.43	- 0.04
	2.8	1.33	0.02	1.50	- 0.06	1.53	- 0.04	1.40	- 0.01
	2.3	1.35	0.00	1.51	- 0.07	1.51	- 0.02	1.36	0.03
	1.8	1.35	0.00	1.49	- 0.05	1.51	- 0.02	1.39	0.00
	1.3	1.38	- 0.03	1.48	- 0.04	1.52	- 0.03	1.41	- 0.02
	0.8	1.38	- 0.03	1.50	- 0.06	1.50	- 0.01	1.46	- 0.07
	0.3	1.39	- 0.04	1.51	- 0.07	1.47	0.02	1.55	- 0.16
Fiber diameter R/T ratio	Whole-tree value (a)	1.19	–	1.14	–	1.28	–	1.21	–
(-)	3.3	1.22	- 0.03	1.20	- 0.06	1.29	- 0.01	1.19	0.02
	2.8	1.22	- 0.03	1.18	- 0.04	1.27	0.01	1.18	0.03
	2.3	1.21	- 0.02	1.16	- 0.02	1.24	0.04	1.18	0.03
	1.8	1.19	0.00	1.19	- 0.05	1.28	0.00	1.20	0.01
	1.3	1.16	0.03	1.21	- 0.07	1.33	- 0.05	1.22	- 0.01
	0.8	1.19	0.00	1.19	- 0.05	1.30	- 0.02	1.25	- 0.04
	0.3	1.22	- 0.03	1.22	- 0.08	1.27	0.01	1.34	- 0.13
Fiber coarseness	Whole-tree value (a)	0.092	–	0.094	–	0.145	–	0.154	–
(mg/m)	3.3	0.096	- 0.004	0.092	0.002	0.147	- 0.002	0.148	0.006
	2.8	0.091	0.001	0.092	0.002	0.143	0.002	0.155	- 0.001
	2.3	0.087	0.005	0.093	0.001	0.140	0.005	0.163	- 0.009
	1.8	0.088	0.004	0.091	0.003	0.142	0.003	0.151	0.003
	1.3	0.089	0.003	0.090	0.004	0.146	- 0.001	0.139	0.015
	0.8	0.094	- 0.002	0.096	- 0.002	0.154	- 0.009	0.153	0.001
	0.3	0.100	- 0.008	0.104	- 0.010	0.158	- 0.013	0.156	- 0.002

References

[1] Downes, G., Hudson, I., Raymond, C., Dean, G., Michell, T., Schimleck, L., Evans, R. and Muneri, A. Sampling plantation eucalyptus for wood and fibre poperties, CSIRO Publications, Melbourne, Australia (1997).
[2] Ona, T., Sonoda, T., Ito, K., Shibata, Y., M., Tamai, Y., Kojima, Y., Ohshima, J., Yokota, S. and Yoshizawa, N. Wood Sci. Technol. 35 (2001) 229.
[3] Hudson, I., Wilson, L., and Van Beveren, K. 54 (2001) 182.
[4] Ona, T., Sonoda, T., Ito, K., Shibata, M., Tamai, Y. and Kojima, Y. Appita J. 49 (1996) 325.

Rapid Assessment of Vessel Morphology by Pyrolysis-Gas Chromatography

Mari Tateishi[a], Teruyuki Seino[b], Kokki Sakai[a], Toshihiro Ona[a], Jyunichi Ohshima[c], Kodai Adachi[c], Shinso Yokota[c] and Nobuo Yoshizawa[c]

[a]*Graduate School of Bioresource and Bioenvironmental Sciences, Kyusyu University, Hakozaki, Higashi-ku, Fukuoka 812-8581, Japan*

[b]*NEDO Fellow, Institute for Environmental Management Technology, AIST Tsukuba West, Tsukuba 305-8569, Japan*

[c]*Faculty of Agriculture, Utsunomiya University, 350 Mine-machi, Utsunomiya 321-8505, Japan*

Introduction

Tree breeding has become more important as fixation sources of atmospheric CO_2, since the global warming has been called a serious environmental problem. Therefore, the selection of elite trees possessing significant CO_2 absorption ability together with excellent pulp properties is strongly needed to accelerate both atmospheric CO_2 absorption and paper recycling rate since stronger papers can keep enough paper strength after repetitive recycling.

Vessel is necessary for the tree growth and its morphology is influenced by the environmental stress such as water availability [1]. Larger vessel areas imply lower tree density and less CO_2 absorption. However, the response magnitude is different among individuals. On the other hand, vessel has a significant influence on the paper properties; when large vessel exists on the paper surface, it is picked from the surface during printing and makes the paper deficient [2]. Furthermore, vessel morphology influences tear strength and stretching of paper [3], and the penetration of pulping liquors into the chip [4]. Here, between-tree variation was genetically observed in vessel morphology [5]. These suggest that vessel morphology could be used as a selection index of elite trees.

Quantitative microscopy is used for conventional vessel morphology determination [6]. However, it takes long time to prepare and analyze samples, and a rapid and easy analytical method is strongly demanded. For this objective, SilviScan [7] and FT-Raman spectrometer [8, 9] are proposed. However, these devices are expensive. On the contrast, a pyrolyzer (Py) is reported to analyze wood samples combined with a widely spread gas chromatograph (GC) [10]. Py-GC is not expensive compared to the above instruments, however; it has never been applied for wood morphological analysis.

In this paper, we examined lignin fragments by using Py-GC to assess vessel morphology.

Materials and Methods

Materials

Two individuals of *Eucalyptus globulus* grown in Western Australia were used.

Conventional method (Quantitative method)

Microtomed cross sections, 15 μm in thickness were prepared from various parts of the trees and were stained with safranine. Anatomical measurements were achieved using a light microscope (Olympus, BH-2) and an image analysis system (Nihon Hybrid, NVS-610). The area and lumen of each 30 vessels were measured at random in a sample, and vessel wall ratio was calculated. The length of 50 vessels macerated with using of Schultz's solution was measured with a digitizer (Oscon, Gradimate SQ-30000) under a light microscope.

Py-GC

A Py-GC system (Frontier Lab, PY-2020S - Shimadzu, GC-17A) with a flame ionization detector (FID) was used. A sample was pyrolyzed at 450°C and a metal capillary column (Frontier Lab, Ultra ALLOY-PY1: 30 m x 0.15 mm i.d x 0.25 μm film thickness) were used. The column temperature was programmed from 50 to 280°C at 4.5°C min^{-1}, and finally held for 10 min. Identification of lignin fragments was carried out by a GC-mass spectrometer (GC-MS) (JEOL, JMS-AMSUN200).

Correlations

Correlation coefficients between lignin fragments and vessel morphology (vessel length, vessel wall thickness and cell wall ratio of vessel) were calculated by the analysis of variance using a statistical computer software, SPSS.

Results and Discussion

Figure 1 shows a typical pyrogram obtained by Py-GC/FID. Various lignin fragments were observed as peaks 1 to 24 on the pyrogram. Table 1 shows the peak assignment of lignin fragments. The relative molar yield of each lignin fragment was determined according to Yokoi et al [10]. Table 2 shows correlation coefficients (<5% level) between lignin fragments and vessel morphology. Table 3 shows correlation coefficients between vessel morphology traits.

As shown in Table 2, we obtained highly significant correlation coefficients between lignin fragments and vessel morphology. Guiacol, peak 1 is suggested as an important index for all traits. This is partly from correlations among traits (Table 3). However, *Eucalyptus* vessel wall is reported as mainly composed of guaiacyl unit [11] and the results obtained are in good agreement with this. We also recognize that the thicker vessel wall shows more *cis*-propenylsyringol and *trans*-sinapyl alcohol. When cell wall becomes thick, more syringyl unit observation is reported [12] and agrees with our results. Speculation of the meaning of lignin fragments related to vessel morphology is difficult since lignin pyrolysis mechanism is not clear. However, the results obtained contain many morphological information, and further studies are encouraged.

Consequently, we showed the feasibility of vessel morphology assessment by Py-GC.

This method will be used for the elite tree selection contributing to help to solve the global warming.

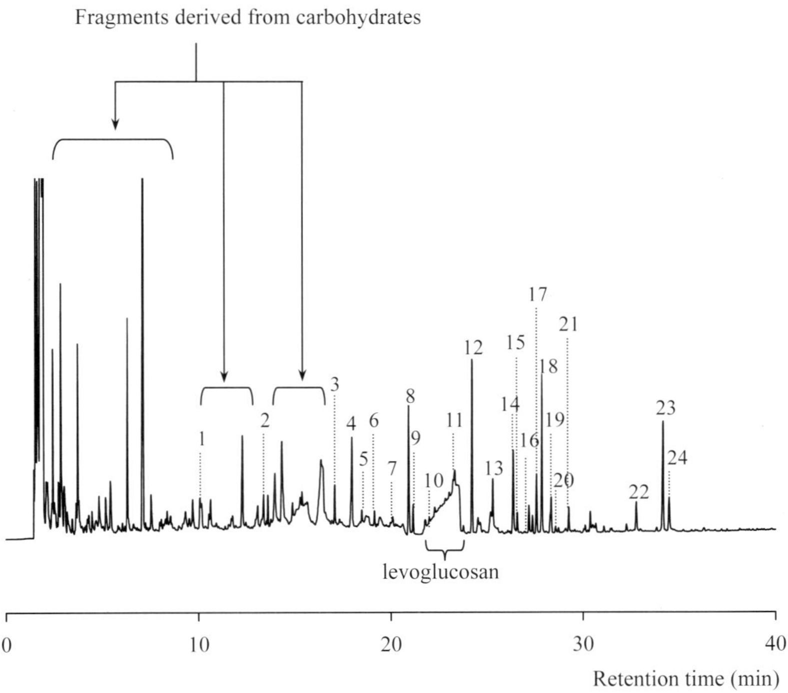

Figure 1 Typical pyrogram detected by FID.

Table 1 Peak assignment of lignin fragment with molar yield. Molar yield is average ± standard deviation.

Peak no.	Lignin fragment	Molar yield (%)	Peak no.	Lignin fragment	Molar yield (%)
1	guaiacol	7.24 ± 0.48	13	allylsyringol	2.39 ± 0.22
2	methylguaiacol	3.80 ± 0.34	14	syringaldehyde	9.44 ± 0.65
3	vinylguaiacol	2.30 ± 0.26	15	*cis*-propenylsyringol	1.07 ± 0.06
4	syringol	10.68 ± 0.59	16	*cis*-coniferyl alcohol	0.12 ± 0.05
5	eugenol	0.75 ± 0.11	17	homosyringaldehyde	4.62 ± 0.33
6	vanillin	1.51 ± 0.18	18	*trans*-propenylsyringol	10.22 ± 0.52
7	*cis*-isoeugenol	0.14 ± 0.04	19	acetosyringone	4.20 ± 0.30
8	methylsyringol	10.52 ± 0.66	20	*trans*-coniferyl alcohol	0.42 ± 0.24
9	*trans*-isoeugenol	1.66 ± 0.18	21	syringylacetone	1.48 ± 0.09
10	acetoguaiacone	0.36 ± 0.10	22	*cis*-sinapyl alcohol	2.58 ± 0.25
11	guaiacylacetone	0.31 ± 0.06	23	*trans*-sinapaldehyde	9.52 ± 0.28
12	vinylsyringol	12.01 ± 0.74	24	*trans*-sinapyl alcohol	2.68 ± 0.94

Table 2 Correlation coefficient between lignin fragment and vessel morphology. VL : vessel length, VWT : vessel wall thickness, VWR : vessel wall ratio.

Peak no.	VL (mm)	VWT(μm)	VWR (%)
1	0.568**	0.531**	- 0.442**
2	0.254	0.317*	- 0.448**
6	0.158	0.389**	- 0.188
13	- 0.188	- 0.398**	0.173
15	0.219	0.344*	- 0.071
16	- 0.189	- 0.278*	- 0.011
19	- 0.228	- 0.360*	- 0.020
23	- 0.125	- 0.277*	0.246
24	0.321*	0.388**	- 0.145

** Significant at 1% level, * Significant at 5% level.

Table 3 Correlation coefficient between vessel morphology.

	VWT (μm)	VWR (%)
VL (mm)	0.728**	- 0.743**
VWT (μm)	—	- 0.600**

** Significant at 1% level, * Significant at 5% level.

References

[1] Cutter, B. E., Murphy, W. K. Wood Fiber 9 (1978) 282.
[2] Colley, J. Paper Technology 14 (1973) 293.
[3] Ogata, Y. Jpn. Tappi 32 (1978) 377.
[4] Hillis, W. E. Appita J. 23 (1969) 89.
[5] McKimm, R. J., Ilic, Y. Aust. For. Res. 17 (1987)19.
[6] Bradbury, S. J. in Light Microscopy in Biology, Lacey, A. J., Ed., IRL Press, Oxford (1989) 187.
[7] Evans, R., Downes, G., Menz, D., Stringer, S. Appita J. 48 (1993) 134.
[8] Ona, T., Sonoda, T., Ito, K., Shibata, M., Tamai, Y., Kojima, Y., Ohshima, J., Yokota, S. Yoshizawa, N. Wood Sci. Technol. 35 (2001) 229.
[9] Ona, T., Sonoda, T., Ito, K., Shibata, M., Ootake, Y., Ohshima, J., Yokota, S., Yoshizawa, N. Analyst 124 (1999) 1477.
[10] Yokoi, H., Ishida, Y., Ohtani, H., Tsuge, S., Sonoda, T., Ona, T. Analyst 124 (1999) 669.
[11] Watanabe, Y., Kojima, Y., Funada, R., Fukazawa, Y., Ona, T., Asada, T. "Recent Advances in Wood Anatomy", Donaldson, L. A., Singh, A. P., Butterfield, B. G., Whitehouse, L. J., Eds., New Zealand Forest Research Institute, Rotorua (1996) 253.
[12] Wu, J., Fukazawa, K and Ohtani, J. Holzforshung 46 (1992) 181.

Assessment of Vessel Anatomical Features in *Eucalyptus camaldulensis* by Pyrolysis-Gas Chromatography

Hironori Haisaki[a], Mari Tateishi[b], Teruyuki Seino[c], Kokki Sakai[b], Toshihiro Ona[b], Jyunichi Ohshima[d], Kodai Adachi[d], Shinso Yokota[d] and Nobuo Yoshizawa[d]

[a]*Faculty of Agriculture, Kyusyu University, Hakozaki, Higashi-ku, Fukuoka 812-8581, Japan*
[b]*Graduate School of Bioresource and Bioenvironmental Sciences, Kyusyu University, Hakozaki, Higashi-ku, Fukuoka 812-8581, Japan*
[c]*NEDO Fellow, Institute for Environmental Management Technology, AIST Tsukuba West, Tsukuba 305-8569, Japan*
[d]*Faculty of Agriculture, Utsunomiya University, 350 Mine-machi, Utsunomiya 321-8505, Japan*

Introduction

Eucalyptus camaldulensis plays an important role as one of the major plantation species for pulpwood especially in sub-tropical regions. Tree selection for high pulp yield and strength in the management of pulpwood forests has attracted interest in the use of plantations with short term rotations to reduce the pulp cost and atmospheric CO_2 [1]. The non-destructive method is preferable in the quality breeding program and the increment core method has an advantage to handle many samples since the core can be drilled out easily from the trunk without felling the tree down [2]. However, the estimation of whole-tree properties by this indirect selection method requires significantly high relationships between wood and pulp properties.

Often vessels have a major effect on causing the paper properties [3-9]; for examples, they are picked from the paper during printing to cause a non-uniform paper surface affecting the distribution of ink [3], they are related to paper properties such as tear strength, paper stretching [4-6], and the penetration of pulping liquors into the wood [7]. These suggest vessel anatomical features could be used as a selection index for quality breeding.

In this paper, we examined the feasibility of the assessment method of vessel anatomical features by pyrolysis-gas chromatography developed using a diffuse-porous species of *Eucalyptus globulus* [10] for a ring-porous species of *Eucalyptus camaldulensis*.

Materials and Methods

Materials

Two individual trees of 14-year-old *E. camaldulensis* were samples at a site in Western Australia as used before [6]. The plantation site annual average temperature and rainfall were 15.5°C and 1,000 mm, respectively. The two sample trees (No.1 and No.2 trees) were selected for average and superior growth characteristics. The tree height and diameter at breast height of samples are 15.2 m and 18.8 cm for a tree No. 1 and 18.1 m and 23.5 cm for a tree No.2, respectively.

Discs of 6 cm thickness were obtained at 1m intervals from 0.3 m above the ground up to a small end diameter of 8 cm after the sample trees were cut down. Two strips of 2 cm width were cut through the center of each disc, and 2 x 2 x 6 cm wood blocks were prepared from pith to bark in one strip. These blocks from one strip were fixed with ethanol and utilized for the measurement of vessel morphology and those from another strip were used for pyrolysis work.

Measurement of vessel anatomical features

Microtomed cross sections of 15 μm thickness were prepared from the block samples and stained with safranine, and then mounted in Bioleit (Ohken Shoji, Tokyo, Japan). The images of the cross-sections were measured using a CCD camera CK3800 (Meiji Techno, Saitama, Japan) coupled to a personal computer and stored as a digital format for image analysis. The vessel morphology of 30 randomly selected cells was measured using the image processing software NIH Image 1.62 (National Institute of Health, Bethesda, MD, USA).

The following vessel anatomical features were measured: radial and tangential diameters, cell area, lumen area, cell perimeter and lumen perimeter. Average diameter and wall thickness were calculated as follows [11]:

Average diameter (μm) = (radial diameter + tangential diameter) / 2

Wall thickness (μm) = 2 x (cell area – lumen area)
/ (cell perimeter + lumen perimeter)

Under these measuring conditions, a dimension of one pixel on the computer corresponded to 0.6 μm of real dimension.

Pyrolysis-gas chromatography

Wood powder samples extracted by a solvent series were ground in a liquid N_2 and were pyrolyzed and analyzed under the same conditions as a previous report [12] using the same analytical systems [10].

Statistical treatement

Correlation coefficients between lignin fragments and vessel anatomical features were calculated by the analysis of variance using a statistical computer software, SPSS.

Results and Discussion

Figure 1 shows a typical chromatogram obtained by Pyrolysis-gas chromatography/flame ionized detector. Various lignin fragments were observed as peaks 1 to 26 on the chromatogram. Table 1 shows the peak assignment of lignin fragments. However, peaks of 8 and 9, and peaks of 20 and 21 were not separated by the same condition except a gas chromatograph hired according to our previous report [12]. Various temperature profiling were examined but the results obtained were not improved. This might be derived from the different performance of a SHIMADZU gas chromatograph utilized instead of an Agilent one.

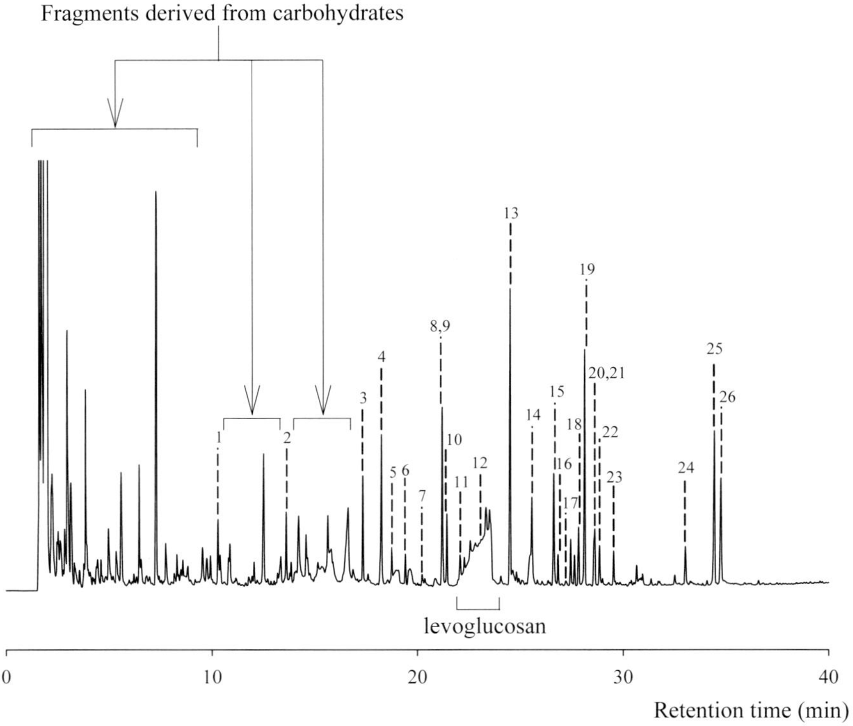

Figure 1 An example of chromatogram of Eucalyptus camaldulensis obtained by pyrolysis-gas chromatography.

Table 2 shows correlation coefficients (over significant at 1% level) between lignin fragments and vessel anatomical features. Table 3 shows correlation coefficients between vessel anatomical features.

As shown in Table 2, we obtained highly significant correlation coefficients between lignin fragments and vessel anatomical features. The important lignin fragments were various structures and the species difference was observed between *Eucalyptus camaldulensis* and *Eucalyptus globulus* [10]. This might be caused by the vessel arrangement difference. Further studies are expected.

The Raman spectroscopy method is very rapid and effective for many wood and pulp properties [13-14]; however, it is affected by the contaminants such as soils. Contrary, the principal of pyrolysis-gas chromatography does not have an influence from them.

We showed the feasibility of vessel anatomical features assessment by pyrolysis-gas chromatography. This method will be used for the elite tree selection for pulp and paper production, especially in the contaminated samples with soils or with decayed parts.

Table 1 Assignment of lignin fragment in Figure 1.

Peak no.	Lignin fragment	Peak no.	Lignin fragment
1	guaiacol	14	allylsyringol
2	methylguaiacol	15	syringaldehyde
3	vinylguaiacol	16	*cis*-propenylsyringol
4	syringol	17	*cis*-coniferyl alcohol
5	eugenol	18	homosyringaldehyde
6	vanillin	19	*trans*-propenylsyringol
7	*cis*-isoeugenol	20	*trans*-coniferaldehyde
8	methylsyringol	21	acetosyringone
9	homovanillin	22	*trans*-coniferyl alcohol
10	*trans*-isoeugenol	23	syringylacetone
11	acetoguaiacone	24	*cis*-sinapyl alcohol
12	guaiacylacetone	25	*trans*-sinapaldehyde
13	vinylsyringol	26	*trans*-sinapyl alcohol

Table 2 Correlation coefficients between lignin fragments and vessel anatomical features. VL: vessel length, VWT: vessel wall thickness, VLDave: vessel lumen average diameter, VDave: vessel diameter average diameter.

Peak no.	VL (mm)	VWT (μm)	VLDave (μm)	VDave (μm)
2	- 0.593**	- 0.427*	-	-
3	- 0.586**	- 0.506**	-	-
5	- 0.451*	-	-	-
6	-	-	0.434**	0.439**
8	- 0.491**	- 0.411*	-	-
13	- 0.373*	- 0.521**	-	-
15	0.593**	0.569**	-	-
22	0.439*	0.419*	-	-
24	0.514**	0.398*	-	-
25	0.471**	0.371*	-	-
26	0.465**	0.387*	-	-

**Significant at 0.1%level, *Significant at 1.0%level

Table 3 Correlation coefficients between vessel anatomical features.

	VWT (μm)	VLDave (μm)	VDave (μm)
VL (mm)	0.628**	0.430*	0.446*
VWT (μm)	—	0.477*	0.504**
VLDave (μm)	—	—	0.999**

**Significant at 0.1%level, *Significant at 1.0%level

References

[1] Ona, T. and Sonoda, T. *Cell. Commun.* 2 (1996) 3.
[2] Downes, G., Hudson, I., Raymond, C., Dean, G., Michell, T., Schimleck, L., Evans, R. and Muneri, A. "Sampling Plantation Eucalypts for wood and fibre properties", CSIRO Publications, Melbourne (1997).
[3] Colley, J. Paper Technology, 14 (1973) 293.
[4] Ogata, Y. Jpn. Tappi J. 32 (1978) 377.
[5] Malan, F. S., Male, J. R. and Venter, J. S. M. Paper Southern Africa 6 (1994) 6.
[6] Ona, T., Sonoda, T., Ito, K., Shibata, M., Tamai, Y., Kojima, Y., Ohshima, J., Yokota, S. and Yoshizawa, N. Wood Sci. Technol. 35 (2001) 229.
[7] Hillis, W. E. Appita J. 23 (1969) 89.
[8] Taylor, F. W. Appita J., 27 (1973) 171.
[9] Bamber, R. K. and Curtin, R. A. Aust. For. 36 (1974) 226.
[10] Tateishi, M., Seino, T., Sakai, K., Ona, T., Ohshima, J., Adachi, K., Yokota, S. and Yoshizawa, N. in this book.
[11] Ohshima, J., Yokota, S., Yoshizawa, N. and Ona, T. Appita J. (2003) in press.
[12] Yokoi, H., Ishida, Y., Ohtani, H., Tsuge, S., Sonoda, T. and Ona, T. Analyst 124 (1999) 669.
[13] Ona, T., Sonoda, T., Ito, K., Shibata, M., Kato, T., Ootake, Y., Tamai, Y. and Kojima, Y. J. Pulp Paper Sci. 26 (2000) 43.
[14] Ona, T., Sonoda, T., Ohshima, J., Yokota, S. and Yoshizawa, N. J. Pulp Paper Sci. 29 (2003) 6.

Chapter 4

Chemical Aspects

Rapid Characterization of Total Fatty Acids in Wood by Reactive Thermal Desorption-Gas Chromatography with Tetrabutylammonium Hydroxide

Miho Mizumoto[a], Teruyuki Seino[b], Kokki Sakai[a], Toshihiro Ona[a], Yasuyuki Ishida[c] and Hajime Ohtani[d]

[a]*Graduate School of Bioresource and Bioenvironmental Sciences, Kyushu University, Hakozaki, Higashi-ku, Fukuoka, Fukuoka 812-8581, Japan*

[b]*NEDO Fellow, Institute for Environmental Management Technology, AIST Tsukuba West, Tsukuba 305-8569, Japan*

[c]*Research Center for Advanced Energy Conversion, Nagoya University, Nagoya 464-8603, Japan*

[d]*Department of Applied Chemistry, Graduate School of Engineering, Nagoya University, Nagoya 464-8603, Japan*

Introduction

Acacia species are widely used for raw materials of papers [1]. However, fatty acid components remaining in paper often cause the paper slide on a printer drum. For this reason, rapid characterization of total fatty acid components in wood is required for the selective tree breeding of *Acacia* species with less fatty acid content.

Conventional method to analyze total fatty acids generally requires tedious and time-consuming sample pretreatments such as solvent extraction, hydrolysis of esterified fatty acids, and derivatization for GC analysis [2]. Furthermore, the fatty acids may be denatured during the pretreatments. On the other hand, a rapid method to characterize fatty acids in organic samples without using the pretreatments was developed on the basis of reactive pyrolysis (Py)-GC with organic alkaline reagents [3, 4]. This technique has been applied to lipid analysis for various samples, such as plant oils, zooplanktons, and cosmetics. Especially, for the analysis of fatty acids in wood, the technique called reactive thermal desorption (RTD)-GC, which performs at lower temperature around 300°C than reactive Py-GC, was developed to achieve the selective thermal desorption and derivatization of fatty acids with suppressing the reactive pyrolysis of major wood constituents, such as lignin and cellulose [5].

For the analysis of fatty acids in wood by RTD-GC, tetramethylammonium hydroxide (TMAH) and tetramethyl ammonium acetate (TMAAc) were used as the typical alkalis agents [5, 6]. However, the total fatty acids analysis included free and esterified ones was not necessarily performed using TMAH or TMAAc; due to the interference of the pyrolyzates from lignin and cellulose for TMAH, and the insufficient hydrolysis efficiency, respectively.

In this study, the compositional analysis of the total fatty acid components in *Acacia* species included free and esterified ones by RTD-GC with has been examined by using another organic alkaline reagent, tetrabutylammonium hydroxide (TBAH).

Experimental

Materials and agents

Three kind of *Acacia* species; *Acacia auriculiformis*, *A. mangium*, and *A. hybrid* (*A. auriculiformis* x *A. mangium*) of age 11 grown in Malaysia were used in this study. The wood samples were cryo-milled into fine powders prior to the RTD-GC measurements. TMAH and TMAAc of 3.0 M aqueous solution and TBAH of 2.6 M aqueous solution were used as the organic alkalis.

Methods

For the RTD-GC procedure, either a GC/mass spectrometer (MS) system (HP6890, Agilent/Automass SUN200, JEOL), or a GC (GC-17A, Shimadzu) equipped with a flame ionization detector (FID) combined with a vertical microfurnance pyrolyzer (PY2020D, Frontier Lab). A metal capillary column (Frontier Lab Ultra ALLOY PY-1, 30 m x 0.25 mm i.d., coated with 0.25 μm of polydimethylsiloxane immobilized through chemical cross-linking) was used as a separation column. About 200 μg of wood samples were taken into a platinum cup and then 3 μl of the organic alkaline solution was added using a micro syringe. The cup was mounted on the waiting position of the pyrolyzer, and then, dropped into the heated center keeping at 300°C to cause RTD under the helium carrier gas. The column temperature was stepwisely programmed from 50°C to 130°C at the rate of 4°C min^{-1} flow, to 165°C at the rate of 2°C min^{-1}, to 210°C at the rate of 1.5°C min^{-1}, to 260°C at the rate of 4°C min^{-1}, to 330°C at the rate of 2.5°C min^{-1}, and held at 330°C for 20 min. Identification of the peaks on the pyrograms was performed mainly by GC/MS.

Results and Discussion

Figure 1 shows chromatograms of *A. auriculiformis* samples obtained by RTD-GC with TMAH (a), TMAAc (b) and TBAH (c), respectively. By using TMAH, RTD of the sample should give the fatty acids methylesters derived from free and esterified ones. However, as shown in Figure 1 (a), the pyrolyzates formed from lignin and other phenolic compounds made the accurate determination of fatty acids very difficult [5]. On the other hand, in the chromatogram of (b), the peaks of fatty acid components were almost missing since the ester bonds in lipids were insufficiently cleaved by TMAAc under the measuring condition used in this work because of the weaker basicity of TMAAc [6]. On the contrary, as shown in the chromatogram (c), RTD-GC of the wood sample in the presence of TBAH gave a series of well-resolved and intense peaks of saturated fatty acids of which butyl esters carbon number was ranging from 16 to 28. This observation suggests that TBAH is enough reactive to cleave the ester bonds in lipids while the decomposition of lignin and cellulose is significantly suppressed during RTD probably because of the slightly weaker basicity of TBAH than TMAH.. Therefore, the RTD-GC with TBAH is for the rapid characterization of total fatty acids in wood.

Figure 2 shows chromatograms of *A. auriculiformis* (a), *A. mangium* (b), and *A. hybrid* (c) obtained by RTD-GC with TBAH. These chromatograms suggest the fatty acids compositions were different to some extent among the three species. For example, the chromatogram of *A. auriculiformis* (a) showed relatively larger peak intensities for C22:0 and C24:0, than that of *A. mangium* (b). These differences can be used to distinguish the two species. On the other hand, the chromatogram (c) doses not

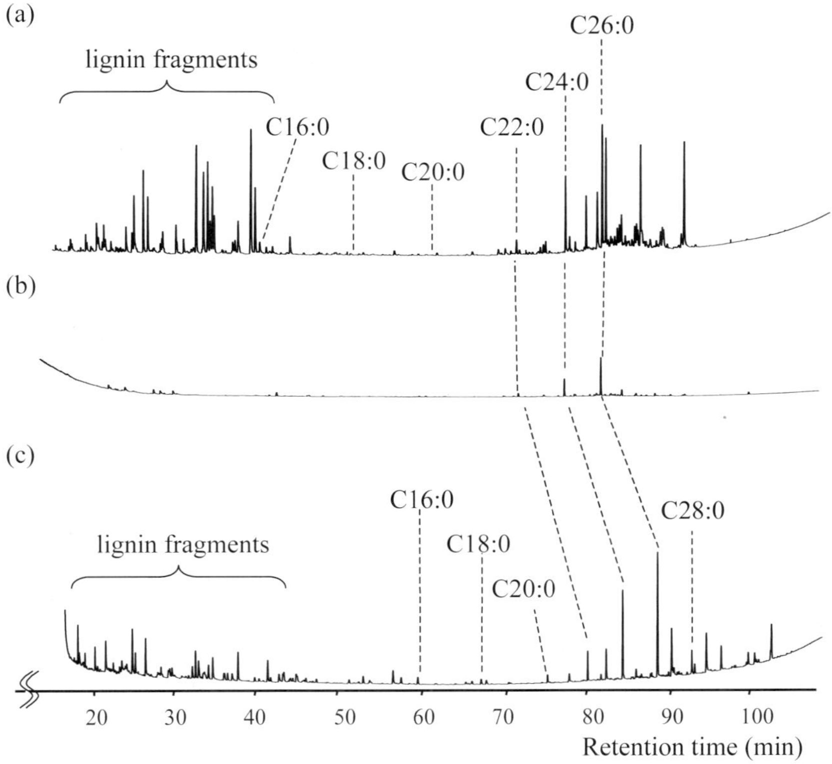

Figure 1 Chromatograms of A. auriculiformis samples obtained by RTD-GC with TMAH (a), TMAAc (b) and TBAH (c).

necessarily show the intermediate pattern between those of the parent species. Furthermore, the chromatograms for *A. hybrid* showed significantly large inter-individual variance. Those results were presumably caused by the difference in the contribution of parent wood though hybridization. In other words, the dominant parent might be judged from the observed chromatograms.

Consequently, the RTD-GC technique with TBAH is effective not only for the compositional analysis of the total fatty acids in *Acacia*, but also for the discriminative analysis among wood species and individuals based on the estimated fatty acid composition. These features could largely contribute for the selective tree breeding of *Acacia* species with less fatty acid content to solve the paper slide problem during printing.

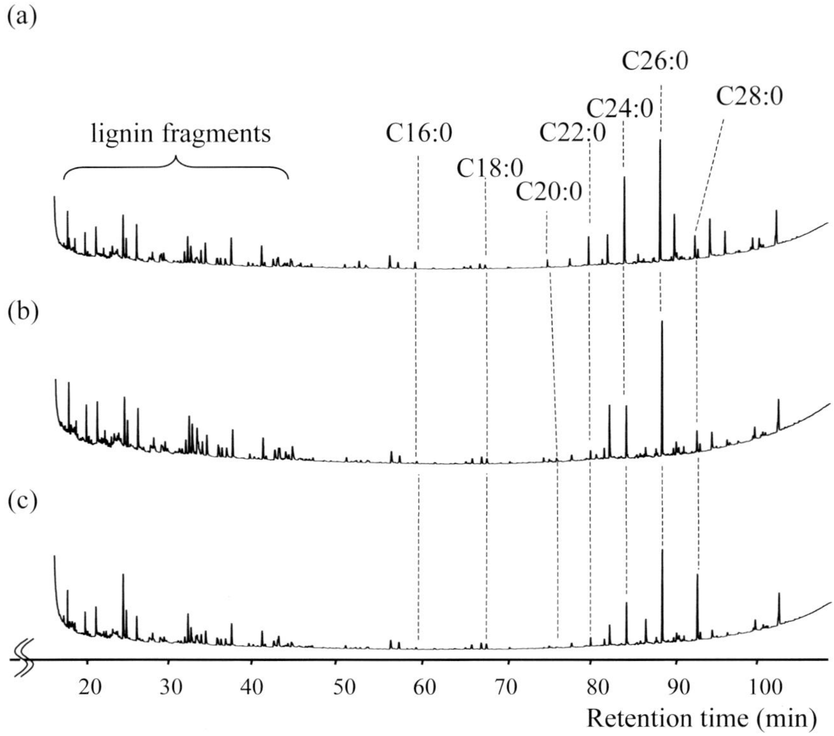

Figure 2 Chromatograms of A. auriculiformis (a), A. mangium (b) and A. hybrid (c) samples obtained by RTD-GC with TBAH.

References

[1] Balodis, V. and Clark, N. B. Appita J. 51 (1998) 179.
[2] Ekman, R. and Holmbom, B. Nord. Pulp Pap. Res. J. 1 (1989) 16.
[3] Challinor, J. M. J. Anal. Appl. Pyrol. 20 (1991) 15.
[4] Challinor, J. M. J. Anal. Appl. Pyrol. 61(2001) 3.
[5] Yokoi, H., Nakase, T., Goto, H., Ishida, Y., Ohtani, H., Tsuge, S., Sonoda, T. and Ona, T. J. Anal. Appl. Pyrol. 67 (2003) 191.
[6] Hardell, H. L. and Nilverant, N. O. J. Anal. Appl. Pyrol. 52 (1999) 1.

Effect of Deuterium Exchange in Lignin on Its Structural Analysis Using FT-Raman Spectroscopy

Shuichi Murakami[a], Toshihiro Ona[b], Kaori Saito[c] and Kazuhiko Fukushima[c]

[a]*Faculty of Agriculture, Kyushu University, Hakozaki, Higashi-ku, Fukuoka 812-8581, Japan*

[b]*Graduate School of Bioresource and Bioenvironmental Sciences, Kyushu University, Hakozaki, Higashi-ku, Fukuoka 812-8581, Japan*

[c]*Graduate School of Bioagricultural Sciences, Nagoya University, Nagoya 464-8601, Japan*

Introduction

In lignin structural analysis, ^{1}H-nuclear magnetic resonance (NMR) spectroscopy is frequently applied using dehydrogenative polymer (DHP) [1]. However, this technique has a disadvantage of long measurement time. On the contrast, Fourier transform (FT) Raman spectroscopy has an advantage of short measurement time and gives the information on various molecule structures similar to NMR. However, peak assignment on the FT-Raman spectrum is difficult because of the peak shift by the surrounding environment for the target structure.

Here, we utilized deuterium exchanged DHP and the effects of deuterium exchanged were examined for the help of peak assignment. In this study, we have used coniferyl alcohol- [γ, γ-$^2H_2$,-OC^2H_3] different from aromatic ring labeling by Agarwal et al [2].

Experimental

The utilized DHP was prepared from labeled or unlabeled coniferyl alcohol was synthesized by the drop method (Figure 1). FT-Raman spectra were collected using a Nicolet Raman 960 spectrometer equipped with an InGaAs detector, and an Nd: YAG laser operating at 1064 nm. Samples were packed into measurement tubes and spectra were collected using 180° backscattering at 300 mW, 4 cm^{-1} resolution and 256 scans at 22°C in a thermal box [3].

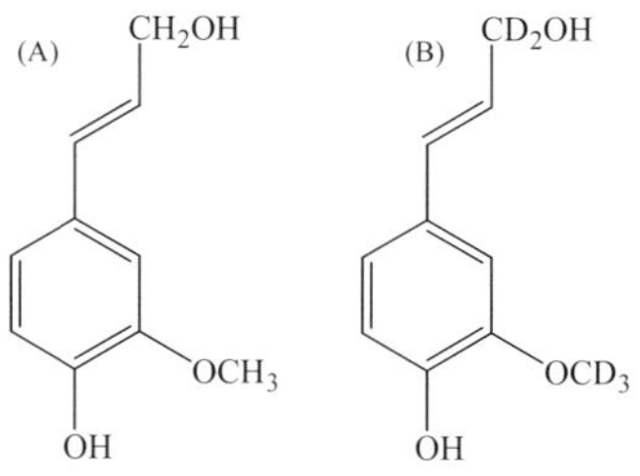

Figure 1 Chemical structure of unlabeled (A) and labeled (B) coniferyl alcohols.

Results and Discussion

Figures 2 and 3 show the FT-Raman spectra of unlabeled and ^{2}H-labeled coniferyl alcohols and corresponding DHPs, respectively. In unlabeled coniferyl alcohol, CH stretching mode from γ-position (primary alcohol) and 3rd position (methoxyl group) are observed at 2940 and 2855 cm^{-1} (Table 1) [4-7].

Table 1 Raman shift (cm^{-1}) and the assignment.

Raman shift (cm^{-1})				Assignments
CA	^{2}H-CA	DHP	^{2}H-DHP	
178	—	—	—	Skeletal deformation
282	—	291	—	Skeletal deformation
379	390	—	—	Skeletal deformation
452	—	—	—	CO deformation from -CH_2OH
552	572	—	—	Aromatic ring deformation
600	—	—	—	OH out-of-plane deformation from -CH_2OH
—	591	—	—	OH out-of-plane deformation from -CD_2OH
738	734	722	726	Aromatic CH out-of-plane deformation
780	761	—	749	Aromatic CH out-of-plane deformation
807	807	811	815	CCO stretch
907	914	—	903	Aromatic CH out-of-plane deformation
969	969	—	965	CH wagging from -HC=CH-
1011	1011	—	—	CCO stretch
1035	—	—	—	?
1085	1077	—	1081	Aromatic CH in-plane deformation
—	—	—	1108	Aromatic CH in-plane deformation
1131	—	1143	1146	Aromatic CH in-plane deformation
1162	1170	—	—	Aromatic CH in-plane deformation
1192	—	1189	—	OCH_3 rocking
1216	1227	—	—	Aromatic CH in-plane deformation
1273	—	1274	—	CH_2 twist from -CH_2OH
1293	1301	1301	1305	CH deformation from -HC=CH-
—	—	1331	1343	CH deformation from α and β position
1373	—	1366	1374	C-C-H bending
1393	—	—	—	CH_2 wagging from -CH_2OH
1451	—	1451	—	CH_2 deformation from -CH_2OH And CH_3 deformation from OCH_3
1602	1605	1602	1602	Aromatic ring stretch
1656	1656	1652	1652	CC stretch from -HC=CH- And CO stretch from aromatic
—	2068	—	2072	CD stretch from -OCD_3
—	2241	—	2261	CD stretch from -CD_2OH
2855	—	2843	—	CH stretch from –OCH_3
2940	—	2936	—	CH stretch from -CH_2OH
—	—	2982	2982	Aliphatic CH stretch
3009	3009	3005	3005	CH stretch from -HC=CH-
3063	3063	3067	3067	Aromatic CH stretch

CA: conifery alcohol, DHP: dehydrogenative polymer.

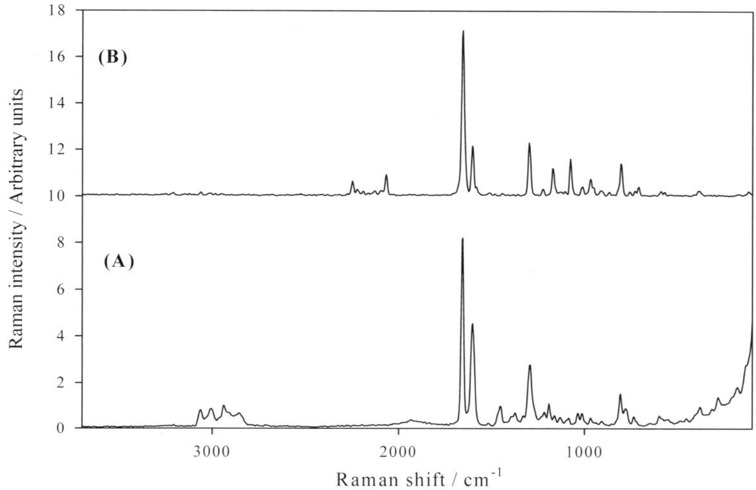

Figure 2 FT-Raman spectra of unlabeled (A) and labeled (B) conifenyl alcohols.

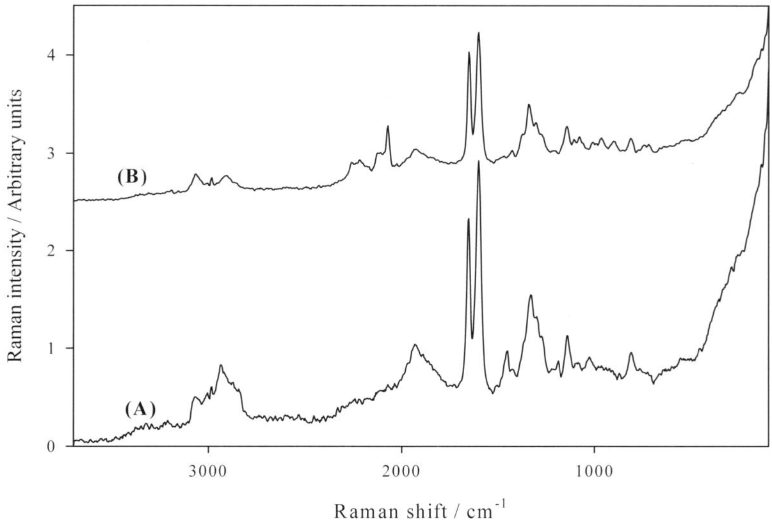

Figure 3 FT-Raman spectra of unlabeled (A) and labeled (B) DHPs.

In labeled coniferyl alcohol, CD stretching mode from γ -position (primary alcohol) and 3rd position (methoxyl group) were observed at 2241 and 2068 cm^{-1} and isotope shifts were recognized [8]. Similarly, isotope shifts were observed in labeled DHP. In unlabeled coniferyl alcohol and corresponding DHP, CH_2 deformation mode and twisting mode from γ position (primary alcohol) and CH_3 deformation mode and rocking mode from 3rd position (methoxyl group) were observed. However, they did not exist in the spectra of labeled coniferyl alcohol and corresponding DHP. This is considered that those vibrations are canceled by the mass effect of deuterium exchange.

Figures 4 and 5 show the ^{1}H-NMR spectra of unlabeled and ^{2}H -labeled coniferyl alcohols, respectively. Figures 6 and 7 show the ^{1}H-NMR spectra of unlabeled and ^{2}H-labeled DHPs, respectively. The characteristic difference was the peak at 4.2 ppm derived from γ -position. They were not observed in ^{2}H -labeled coniferyl alcohol and corresponding DHP. Furthermore, the peak at 3.9 ppm in unlabeled coniferyl alcohol and corresponding DHP are assigned to 3rd position from benzene. They did not exist in labeled spectra.

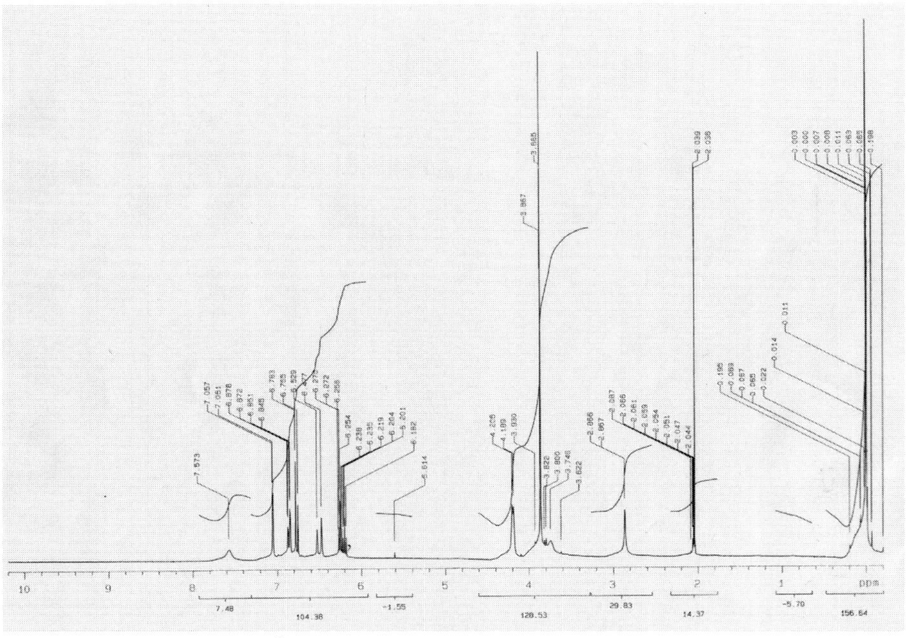

Figure 4 ^{1}H-NMR spectra of unlabeled coniferyl alcohol.

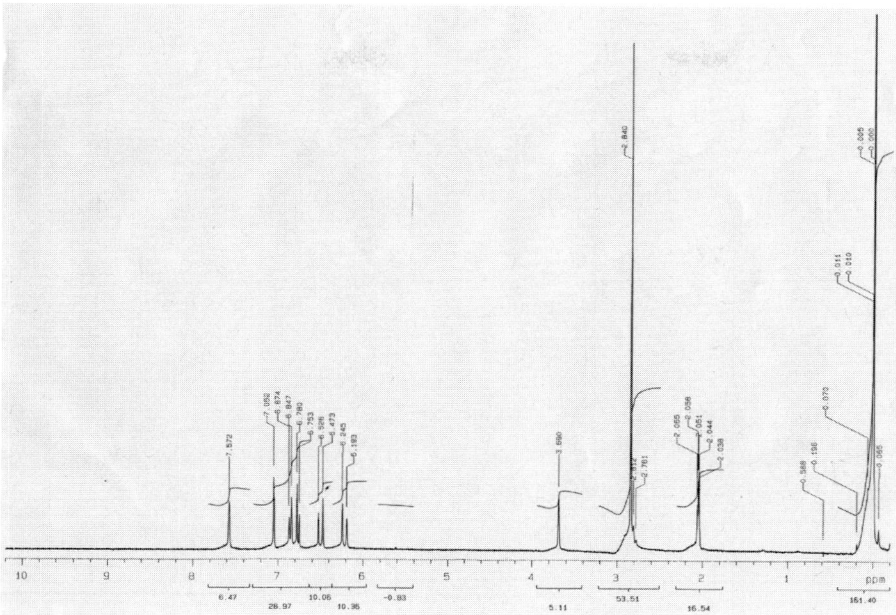

Figure 5 ^{1}*H-NMR spectra of labeled coniferyl alcohol.*

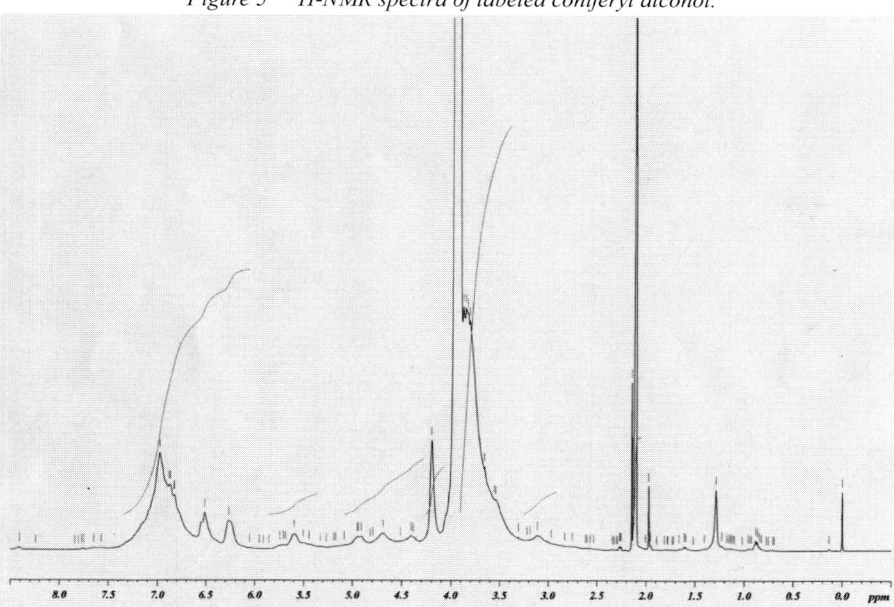

Figure 6 ^{1}*H-NMR spectra of unlabeled DHP.*

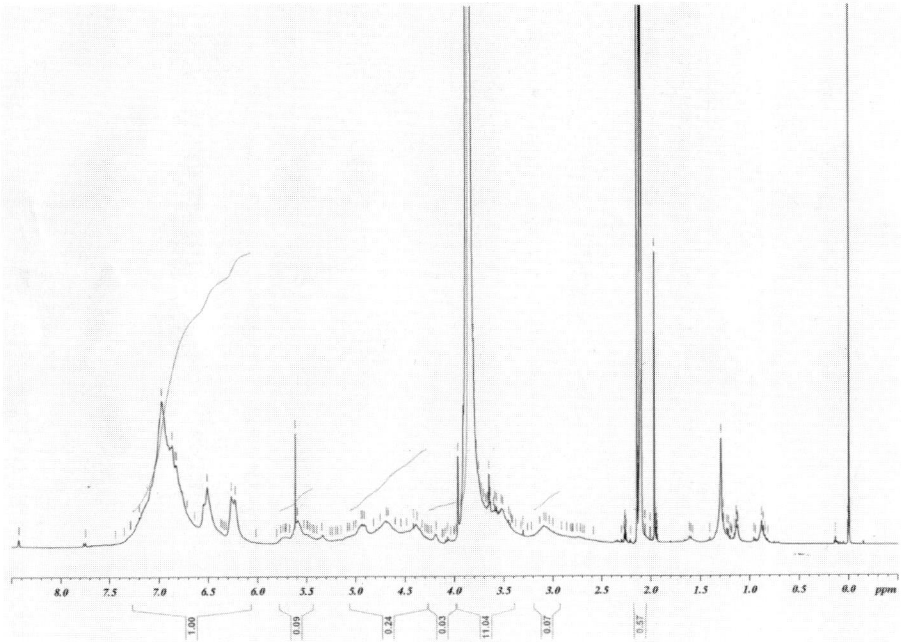

Figure 7 ¹H-NMR spectra of labeled DHP.

Consequently, the results obtained by FT-Raman spectroscopy agree with the disappearance of the peak by the deuterium exchange detected by ¹H-NMR spectroscopy. The use of deuterium exchanged DHP for Raman spectroscopy will contribute to perceive the lignin structure by the peak shift in a Raman spectrum. The data deposition will promise the rapid analysis of lignin structure using FT-Raman spectroscopy.

References

[1] Faix, V. O. and Schweers, W. Holzforschung 28 (1974) 94.
[2] Agarwal, U. P. and Terashima, N. Proceedings of 12th International Symposium on Wood and Pulping Chemistry Vol. III, June 9-12, 2003, Madison, USA (2003) 123.
[3] Ito, K., Kato, T. and Ona, T. J. Raman Spectrosc.32 (2001) 389.
[4] Evans, P. A. Spectrochim. Acta 47A (1991) 1441.
[5] Ona, T., Sonoda, T., Ito, K., Shibata, M., Ootake, Y., Ohshima, J., Yokota, S. and Yoshizawa, N. Anal.Biochem.268 (1999) 43.
[6] Ona, T., Sonoda, T., Ito, K., Shibata, M., Kato, T., Ootake, Y., Tamai, Y. and Kojima, Y. J. Pulp Paper Sci. 26 (2000) 43.
[7] Socrates, G. "Infrared and Raman characteristic group frequencies: tables and charts (3rd ed.)", John Wiley & Sons, Chichester, UK (2001).
[8] Pawlukojc, A., Bator, G., Sobczyk, L., Grech, E. and Nowicka-Scheibe, J. J. Phys. Org. Chem. 16 (2003) 709.

Direct Analysis of Condensed Tannins in Bark by Matrix-Assisted Laser Desorption/Ionization Mass Spectrometry Combined with Solid Sample Preparation

Yasuyuki Ishida[a], Kuniyoshi Goto[b], Hajime Ohtani[b] and Tetsuya Sonoda[c]

[a]Research Center for Advanced Energy Conversion, Nagoya University, Furo-cho, Chikusa-ku, Nagoya 464-8603, Japan

[b]Graduate School of Engineering, Nagoya University, Furo-cho, Chikusa-ku, Nagoya 464-8603, Japan

[c]Kameyama Research Center, Forestry Research Institute, Oji Paper, 24-9 Nobono-cho, Kameyama 519-0212, Japan

Introduction

Condensed tannins, mainly comprised of flavan-3-ol (catechin and epicatechin) units, are one of the phenolic compounds in wood with molecular weights between 500-3000 as shown in Figure 1 [1, 2]. The condensed tannins have been traditionally utilized as the agents for converting animal hides to leather (tanning). Recently, the tannins have been also used as an adhesive for wood materials, coagulant and coating material, and expected to be utilized as pharmaceutical products because of their biological activities [3].

The conventional methods for the characterization of tannin components usually involve their solvent extraction from wood samples followed by liquid chromatographic separation. The solvent extraction process, however, not only is tedious and time-consuming but also often leads to insufficient recovery. In this work, a new method to analyze condensed tannin components in wood directly without using solvent extraction was developed by matrix-assisted laser desorption/ionization mass spectrometry (MALDI-MS) combined with a solid sampling technique [4-7].

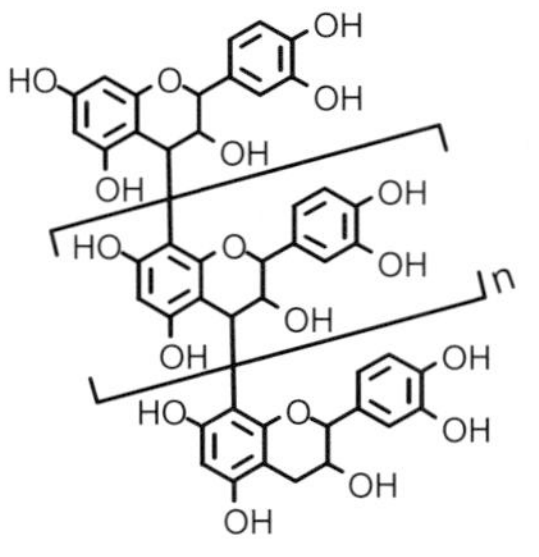

Figure 1 Typical structure of condensed tannins [1, 2].

Experimental

The condensed tannins are known to be abundantly contained in the bark of timber species, such as conifers and eucalyptus trees, at the level of a few tenth of %. Therefore, bark of *Acacia auriculiformis* was used as the wood sample in this work.

2,5-dihydroxybenzoic acid (DHB) and NaI were used as the matrix and cationization reagents, respectively, for MALDI-MS measurements. An aliquot of the cryo-milled wood powder sample (c.a. 1 mg) was suspended in 500 μl of DHB (125 mM) and NaI (1 mM) solution in methanol containing 1% (v/v) trifluoroacetic acid (TFA). Here, TFA was added into the matrix solution in order to enhance the ion formation of the tannin components during MALDI in the wood sample, which might promote the decomposition of their cell wall. After vortexed for 1 min, c.a. 1 μl of the sample suspension was deposited on a flat stainless steel MALDI plate and allowed to air-dry prior to analysis. MALDI-MS measurements were performed using Voyager-DE RP (Applied Biosystems, Inc.) equipped with a pulsed nitrogen laser (337 nm).

MALDI-mass spectrum of the extractives conventionally solvent-extracted from the wood sample was also obtained by an ordinary solution-based sampling technique (dried droplet method). At first about 5 g of the powdered wood sample was subjected to Soxhlet extraction for 12 h using 300 ml of methanol/water (v/v = 8/2). The condensed extracts using a rotary evaporator were subsequently redissolved to prepare 60 ml of the methanol/water solution. Then equal volumes of the sample solution and the matrix solution of DHB (250 mM) and NaI (2 mM) in methanol were blended, and vortexed for 1 min. After the addition of 1 % (v/v) TFA, 1μl of the mixed sample solution was deposited on a MALDI plate and finally subjected to MALDI by laser irradiation.

Results and Discussion

Figure 2 shows typical MALDI mass spectra of (a) tannin components directly desorbed from the wood sample by the solid sampling technique and (b) extracts from the wood sample observed by an ordinary solution-based sample preparation. The spectrum (a) clearly shows a series of ion groups consisting of a few peaks for each group up to around m/z 3,000. The most intense peak for each group is separated one another by 288 Da, which is corresponding to the mass of a flavan-3-ol unit. Furthermore, the peaks in each group appeared with intervals of 16 Da. The main peak in each group, e.g. m/z 1179 and 1467 in the expanded mass spectrum, can be attributed to sodium adduct ions of the typical condensed tannin structure shown in Figure 1 on the basis of their masses. The other peaks in each group are estimated to be their analogues, which differ in the number of hydroxyl groups contained in the molecules. These results suggest that the condensed tannins in the bark of *Acacia auriculiformis* sample used in this work consist of the polymerized flavan-3-ol (shown in Fig. 1) and their analogues at least up to 10-mers (m/z ≈ 3,000). On the contrary, in the spectrum of the extracts from the wood sample, the tannin components around 5 and 6-mers are observed as considerably weaker peaks owing to their insufficient recoveries. Moreover, the oligomers higher than 7-mers are almost missing, suggesting that such larger molecules are difficult to be solvent-extracted.

As shown above, MALDI-MS combined with a solid sampling technique is a powerful method to analyze condensed tannins in wood directly without using any troublesome sample pretreatments even for the larger components, which are unsuitable for the conventional solvent extraction. This is an promising method for a rapid and convenient screening of wood by focusing on the amount and distribution of tannin components.

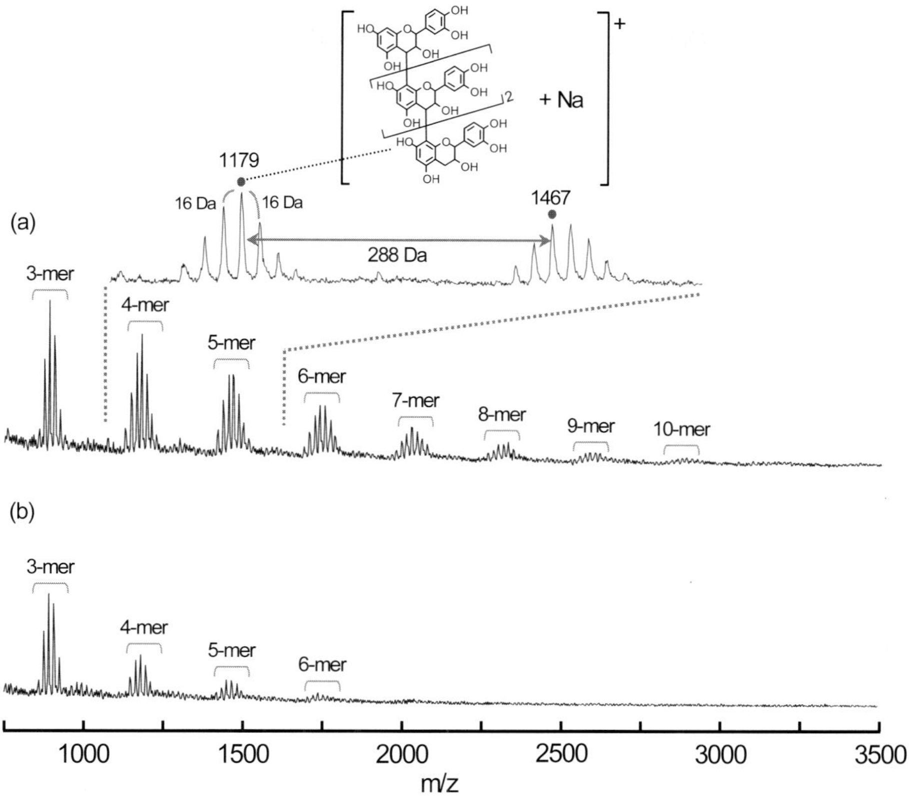

Figure 2 MALDI mass spectra of (a) condensed tannin components directly desorbed from wood sample by a solid sampling technique and (b) extracts from wood sample observed by a solution-based sample preparation.

Acknowledgments

This work was supported in part by the 21st Century COE Program "Nature-Guided Materials Processing" of the Ministry of Education, Culture, Sports, Science and Technology.

References

[1] Pizzi, A. and Stephanou, A. J. Appl. Polym. Sci. 51 (1994) 2109.
[2] Pasch, H., Pizzi, A. and Rode, K. Polymer 42 (2001) 7531.
[3] Santos-Buelga, C. and Scalbert, A. J. Sci. Food. Agric. 80 (2000) 1094.
[4] Skelton, R., Dubois, F. and Zenobi, R. Anal. Chem. 72 (2000) 1706.
[5] Marie, A., Fournier, F. and Tabet, J. C. Anal. Chem. 72 (2000) 5106.

[6] Trimpin, S., Rouhanipour, A., Az, R., Rader, H. J. and Mullen, K. Rapid Commun. Mass Spectrom. 15 (2001) 1364.
[7] Leuninger, J., Trimpin, S., Rader, H. J. and Mullen, K. Macromol. Chem. Phys. 202 (2001) 2832.

Tannin and Related Polyphenolic Components from Tropical Tree Species

Seiji Ohara[a], Koh Hashida[a], Rei Makino[a] and Sulaeman Yusuf[b]
[a]*Department of Forest Chemistry, Forestry and Forest Products Research Institute, 1 Matsunosato, Tsukuba 305-8687, Japan*
[b]*Research Development Centre for Applied Physics, Indonesia Institute of Sciences, Komplek PUSPIPTEK, Serpong 15314, Indonesia*

Introduction

Tannins are widely distributed in tree species such as conifers, eucalypts and legumes. They are divided into two major groups, condensed tannins and hydrolysable tannins. It is well known that condensed tannins undergo some condensation reactions in alkaline media [1,2]. Thus, it is supposed that condensed tannins partly recondense with lignin fragments during kraft pulping process and consequently affect the pulp properties on recycling uses. It is also well known that tannins are present at very high concentrations in the bark of timber species. Bark is one of the main unused resources, and large quantities of barks are burned up in Japan [3]. Because the regulation of incinerators was made strict in 2002 due to the problem of dioxins, the development of utilization of bark has become a very important issue. In this study, characterization of tannin and related polyphenolic components from the wood and bark of tropical tree species (*Eucalyptus camaldulensis*, *Acacia mangium* and *A. auriculiformis)* was conducted in order to obtain the basic information relating to above matters.

Methods

E. camaldulensis woods (2- and 40-year-old trees) were extracted with 80% MeOH or 70% acetone aqueous solution, and polyphenols and flavanols were determined according to Folin-Ciocalteu and vanillin-HCl methods, respectively. Furthermore, polyphenolic compounds from the wood of 40-year-old *E. camaldulensis* were fractionated and purified by Sephadex LH20 column chromatography using ethanol or 50% methanol as an eluent. Isolated compounds and purified condensed tannin polymers were analysed by NMR, FAB-MS, CD and pyrolysis-gas chromatography/mass spectrometry (Py-GC/MS). Tannins and related polyphenolic components from *A. mangium* and *A. auriculiformis* were also analysed according to the same procedures as above.

Results and Discussion

Table shows the results of quantitative analysis of 2- and 40-year-old *E. camaldulensis* woods. Extracts with 70% acetone aqueous solution contained more extractives, polyphenols and flavanols than those with 80% MeOH aqueous solution, suggesting that 70% acetone aqueous is a more effective solvent in extracting polyphenols and tannins. As for the difference of the tree age, the 40-year-old wood contained higher amounts than the 2-year-old wood whichever solvents were used. From the wood of 40-year-old tree, several kinds of polyphenolic components were identified (Figure 1). Of these compounds, catechin and 3, 4, 3', 4' - tetrahydroxy-stilbene were major components in the wood. Polyphenol contents of the bark of *A. mangium* and *A. auriculiformis* were 17.7 and 14.8%, respectively. On the other hand, the contents of polyphenols of the wood of both species were very low (below 1.0%). From the bark of *A. auriculiformis*, gallocatechin, epigallocatechin, epirobinetinidol - (4β-8) - epigallocatechin and 7, 3', 4', 5' - tetrahydroxy-flavan - (4β-8) - epigallocatechin were identified. 7, 3' ,4', 5' - tetrahydroxy-flavan - (4β-8) - epigallocatechin was a novel compound, the NMR spectrum of which showed no rotational isomer probably due to the absence of hydroxyl group at C-3 position.

Table Quantitative analyses of 2- and 40-year-old E. camaldulensis woods.

	2-year-old (%)		40-year-old (%)	
	80% MeOH	70% acetone	80% MeOH	70% acetone
Extractives	5.49	6.66	7.73	8.56
Polyphenols[a)]	0.99	1.30	2.02	2.27
Flavanols[b)]	2.21	3.38	4.05	4.55

Values are % on oven-dried wood meals. [a)]Folin-Ciocalteu method, [b)]Vanillin-HCl method.

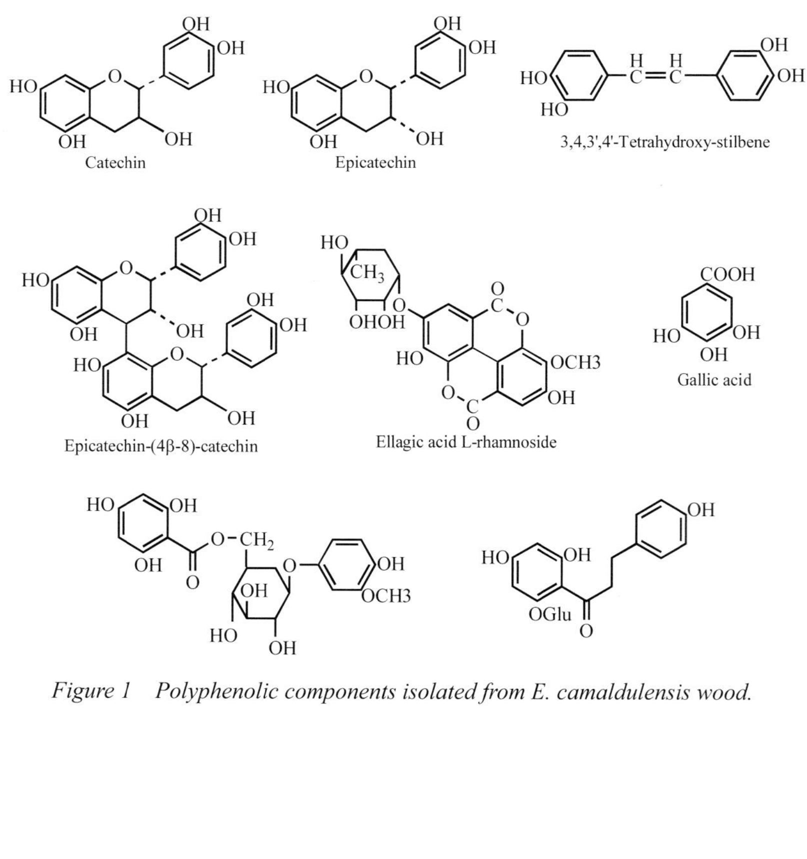

Figure 1 Polyphenolic components isolated from E. camaldulensis wood.

The ^{13}C-NMR spectrum of condensed tannin polymer purified from the wood of *E. camaldulensis* showed broad peaks, but it clearly indicated the homogeneity of the flavan B-ring. The signals at 116 ppm and 120 ppm were assigned to unsubstituted catechol ring carbons, and the signal at 108 ppm due to unsubstituted pyrogallol ring carbons was not detected. Examination of the chemical shifts in the heterocyclic region showed resonances at 76 ppm and 72 ppm corresponding to C-2 and C-3 chemical shifts of flavans with 2, 3 - *cis* configuration and those at 37 ppm corresponding to C-4 of extender units with phloroglucinol type A-ring [4]. The pyrogam of condensed tannin polymer purified from the wood of *E. camaldulensis* is shown in Figure 2a. Catechol was the main pyrolysis product, and reatively much smaller amounts of 4-methylcatechol was also detected. These results indicate that condensed tannins from *E. camaldulensis* consist mainly of catechol type B-ring [5]. Figure 2b shows the pyrogram of condensed tannin polymer from the bark of *A. auriculiformis*. Resorcinol, pyrogallol and 5-methylpyrogallol were observed, but no catechol and 4-methylcatechol. Therefore, it is assumed that condensed tannins from *A. auriculiformis* consist mainly of pyrogallol type B-ring.

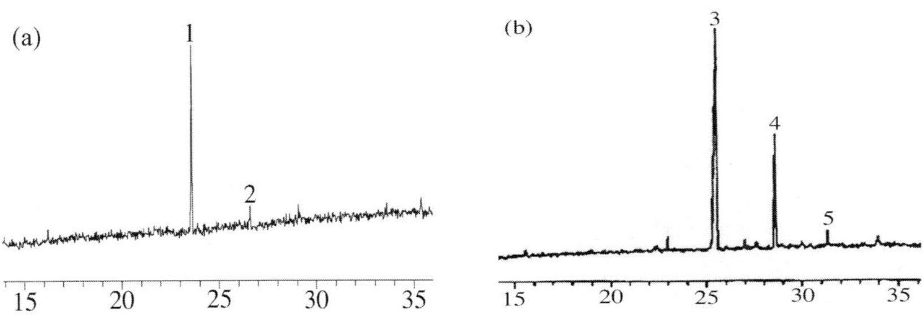

Figure 2 Pyrograms of tannin polymers from the wood of E. camaldulensis (a) and bark of A.auriculiformis (b). 1: Catechol, 2: 4-Methylcatechol, 3: Resorcinol, 4: Pyrogallol, 5-Methylpyrogallol.

References

[1] Ohara, S. and Hemingway, R.W. J. Wood Chem. Technol. 11 (1991) 195.
[2] Hashida, K. and Ohara, S. J. Wood Chem. Technol. (2003) *in press*.
[3] Ikami, Y. and Murata, K. Bulletin of FFPRI 2 (2003) 11.
[4] Ohara, S., Ohira, T. and Suzuki, K. Mokuzai Gakkaishi 40 (1994) 1363.
[5] Ohara, S., Yasuta, Y. and Ohi, H. Holzforschung 57 (2003) 145.

Characterization of Natural Resin Shellac by Reactive Pyrolysis-Gas Chromatography

Lili Wang [a,*], Yasuyuki Ishida [a], Hajime Ohtani [a,†], Shin Tsuge [a] and Toshihiro Nakayama [b]

[a]*Graduate School of Engineering, Nagoya University, Furo-cho, Chikusa-ku, Nagoya 464-8603, Japan*
[b]*Gifu Shellac MFG. Co., Ltd., Gifu 500-8618, Japan*

**Present address: School of Chemical Engineering and Material, Zhejiang University of Technology, Hangzhou 310014, China*

[†]*To whom all correspondence should be addressed.*

Introduction

Shellac is one of the thermosetting resins of animal origin secreted by the lac insect *Kerria lacca* which parasitically grows on some types of trees such as mulberries commonly found in India, Thailand, Myanmar and south area of China. Shellac has been used as thermoplastics, adhesives and sealants, insulating materials and coating materials in various fields such as industrial material, medicine and food ingredient due to its various unique properties such as thermoplasticity, oil-resistibility, cohesiveness and insulating ability along with its non-poisonous nature [1]. Shellac is provided as a refined form of the stic-lac resin (raw shellac, just collected from twigs of the trees) consisting of resin (70-80%), wax (6-7%), coloring matter (4-8%) and others(15-25%) such as debris and moisture. The resin portion constituting the backbone of shellac is a complex mixture of polyesters consisting of a number of closely related sesquiterpenic acids of the cedrene skeleton, mainly jararic acid and laccijalaric acid, and hydroxy-fatty acids, mainly threo-aleuritic acid, which can be separated into about 30% of soft resin (single ester) and about 70% of hard resin (polyesters consisting of several resin acid components) [1]. The chemical composition of shellac which affects its mechanical and thermal properties often varies to some extent depending on nature of the host trees on which the insect grows, the species of the insect and the environmental conditions. Therefore, it is important to know the subtle differences in the chemical composition of shellac.

Various analytical techniques such as chemical analysis, and chromatographic and spectrometric methods have been applied to semi-quantitative determination of shellac resin. In these methods, however, cumbersome pretreatments such as chemical decomposition, derivatization and/or solvent extraction were inevitably utilized to separate the resin constituents prior to the final analytical measurements. Recently, Py-GC/MS in the presence of an organic alkali, tetramethylammonium hydroxide ($(CH_3)_4NOH$, TMAH) has been successfully applied to the characterization of small amounts of natural products consisting of various terpenic acids [2, 3] without using any tedious pretreatments.

In this work [4], reactive Py-GC in the presence of TMAH was applied to the compositional analysis of natural resin shellac. Furthermore, discriminative analysis among shellac samples produced in India and Thailand was carried out by means of the principal component analysis (PAC) on the basis of the chemical composition of all the resin acid components determined by Py-GC.

Materials and Methods

Materials

Eight shellac samples formed from stick-lac gathered in India and Thailand were used after purification. All the shellac samples were cryo-milled into a fine powder (less then 60 mesh) by a freezer mill (SPEX 6700) prior to Py-GC measurements in order to improve the efficiency of the reactive pyrolysis. An aqueous solution (25% wt/wt) of tetramethylammonium hydroxide (TMAH), supplied by Tama Chemicals (Tokyo, Japan), were used as the regent for the reactive pyrolysis.

Reactive Py-GC Condition

A vertical microfurnace pyrolyzer (GP1018, Yanaco, Kyoto, Japan) was directly attached to a gas chromatograph (HP 5890, Hewlett-Packard, Avondale, PA, USA) equipped with a flame ionization detector (FID). About 90 μg of shellac sample and 2 μl of the TMAH solution taken in a platinum sample cup were introduced into the heated center of the pyrolyzer under the flow of helium carrier gas. The relatively low pyrolysis temperature of 400°C was used to suppress the competitive ordinary pyrolysis at higher temperatures. A fused silica capillary column (Hewlett-Packard Ultra 1, 25 m × 0.2 mm i.d.) coated with polydimethylsiloxane (0.33 μm film thickness), immobilized by chemical cross-linking was used. The flow rate of 50 ml/min of carrier gas at the pyrolyzer was reduced to 1.0 ml/min at the capillary column by means of a splitter. Identification of the peaks on the resulting pyrograms was carried out by use of a GC/MS system (JEOL AM-II 150) to which the same pyrolyzer was also attached

Results and Discussion

Figure 1 shows typical pyrograms of Indian shellac at 400°C, obtained without adding TMAH (a) and in the presence of TMAH aqueous solution (b). In the pyrogram of Figure 1 (a), only a series of very weak and broad peaks of terpenic acids and wax components were observed. Furthermore, no peak of the hydroxy fatty acids such as aleuritic acid was observed mainly due to their extremely high polarity. On the contrary, a series of sharp peaks due to methyl derivatives of shellac constituents were observed in Figure 1 (b) with far better resolution and higher sensitivity after the elution of the reagent related peaks. Figure 2 shows a general scheme of reactive pyrolysis for the main components of shellac resin in the presence of TMAH. As shown in this figure, methyl derivatives of shellac constituents were formed as a result of the selective hydrolysis of ester bonding followed by simultaneous methylation of the carbonyl and hydroxyl groups. Moreover, not only characteristic products of the terpenic acids, such as laccijalaric and jalaric acids, and aleuritic acids, but also various minor C_{14}-C_{16} straight chain fatty acids, hydroxy-fatty acids and wax components such as C_{28}-C_{34} even-numbered straight-chain alcohols were also observed as their methyl derivatives.

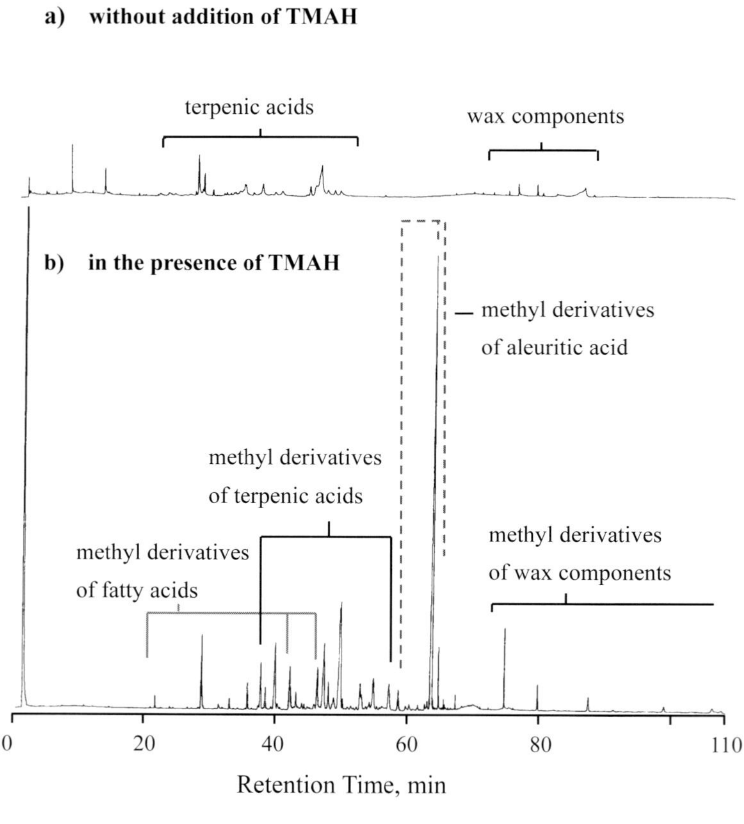

Figure 1 Typical pyrograms of Indian shellac at 400°C without addition of TMAH (a) and in the presence of TMAH aqueous solution (b).

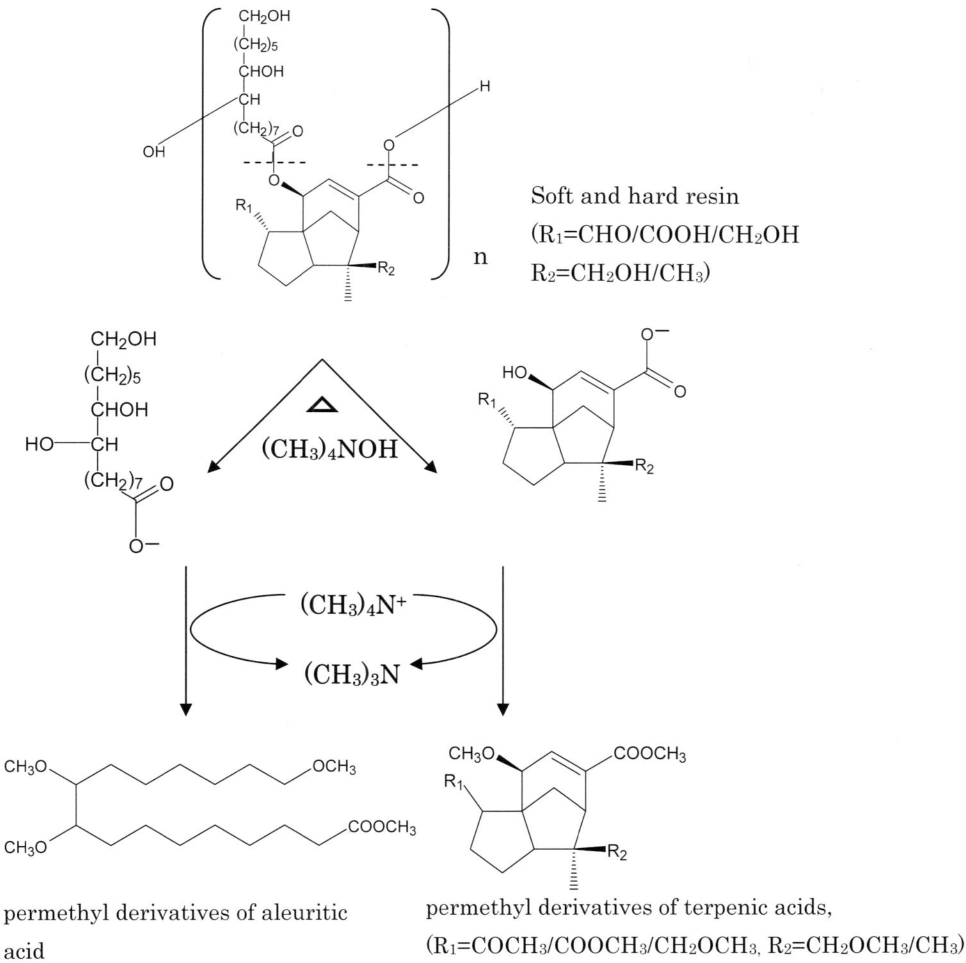

Figure 2 General scheme of reactive pyrolysis for the main components of shellac resin in the presence of TMAH.

Chemical composition of resin acid for each shellac sample was determined on the basis of the peak intensities on the pyrograms obtained during the reactive pyrolysis in the presence of TMAH aqueous solution. The distributions of the total terpenic acids, aleuritic acid and the other fatty acids are fairly similar among the eight samples regardless of the difference in their origin. On the other hand, the relative amounts of several constituents of terpenic acids such as laccijalaric and jalaric acids, and minor constituents of aliphatic and hydroxy-fatty acids proved to be different to some extent between Indian and Thailand shellac.

In order to distinguish the shellac samples of different origin clearly, the PCA method was applied in which the obtained distributions of all the acidic components (14 components) for eight shellac samples produced in India and Thailand were used as the data base after converting them into the deviation values. The contributory rates (dispersion) of the first, second and third principal components obtained were 85.6, 6.5 and 4.3%, respectively. Figure 3 shows the relationship between the first and second principal component scores for eight shellac samples from India and Thailand. As shown here, these shellac samples are clearly divided into two different groups according to their origin.

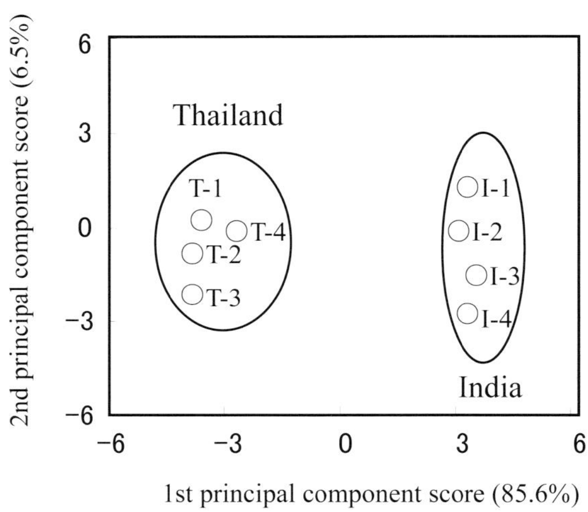

Figure 3 Discriminative analysis among shellac samples produced in different country.

In conclusion, reactive Py-GC in the presence of organic alkali proved to be a rapid, convenient, and highly sensitive method to determine the chemical composition of a natural resin shellac without using any cumbersome pretreatments. Furthermore, principal component analysis for the distribution of the all the acidic components determined can clearly discriminate shellac samples from the two different growing places.

Reference

[1] Maiti, S. and Rahman, M. S. Macromol. Chem. Phys. C26 (1986) 441.
[2] Anderson, K. B. and Winans, R. E. Anal. Chem. 63 (1991) 2901.
[3] Pastorova, I., Van der Berg, K. J., Boon, J. J. and Verhoeven, J.W. J. Anal. Appl. Pyrolysis 43 (1997) 41.
[4] Wang, L., Ishida, Y., Ohtani, H., Tsuge, S. and Nakayama, T. Anal. Chem. 71 (1999) 131.

Conversion Pattern of Lignocellulosics in the Phase-Separation System

Keigo Mikame and Masamitsu Funaoka
Faculty of Bioresources, Mie University, 1515 Kamihama-cho, Tsu 514-8507, Japan

Introduction

In recent years, considerable efforts have been devoted to utilize biomaterials placing petroleum for chemical feedstock. It is of importance to produce biodegradable materials from renewable resources such as lignocellulosics. In this concern, wood is the most important resource because wood can build up large storage of carbon dioxide. However, the wood refining must be done before the use in chemical feedstock. Separating lignocellulosic components is extremely difficult since the lignocellulosics have the interpenetrating polymer network structure composed of lignin, hemicellulose and cellulose [1]. So far the lignocellulosics have been digested with strong alkaline agents for paper making or pre-hydrolyzed with strong acid for ethanol fermentation. During these treatments, lignin are subjected to very complicate modifications [2-4] and deactivated to be treated as a waste.

Recently we have developed a new process for separating and converting lignocellulosics into lignin and carbohydrates in highly reactive forms [5]. This process includes a phase-separation reaction composed of phenol derivatization and concentrated acid, which are immiscible with each other. In the process the concentrated acid is a solvent for carbohydrates, but also works for the fragmentation and phenolation of lignin as a catalyst, whereas phenol derivatives simultaneously act as phenolation agents, a barrier to minimize the attack of acid on lignin and a solvent for the lignin fractions (Figure 1).

The separation of lignocellulosic components is achieved almost quantitatively at

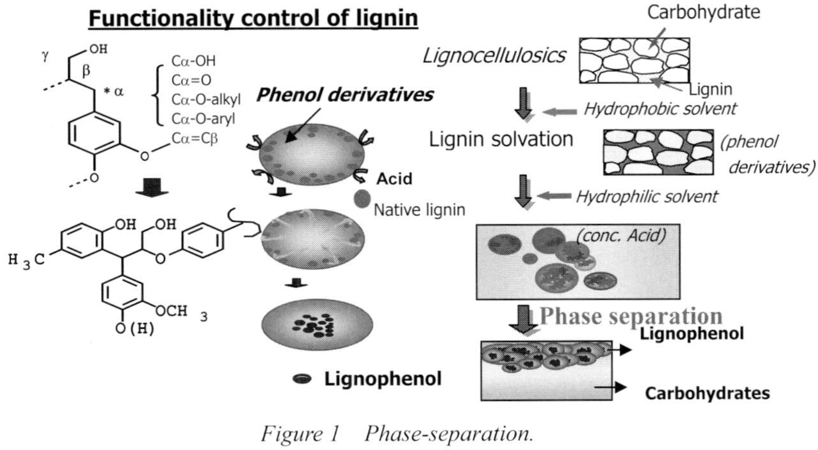

Figure 1 Phase-separation.

room temperature regardless to species. The separated lignin derivatives (lignophenol) have several unique functions discriminated form conventional lignin with retention of the original interunit linkages: these include properties of highly phenolic, no conjugated, lighter color compared to conventional lignin, solid-liquid transformation and high immobilization capability for proteins (enzymes) [6, 7]. Furthermore, the molecular structure of lignophenol can be secondarily controlled using a nucleophilic attack of phenolic units to adjacent carbons, resulting in alteration of important functions such as phenolic activity and molecular weight. These original functions are due to the selective hybridization of lignin structural units with monomeric phenol derivatives at C-1 positions, leading to liner type structure composed mainly of 1, 1-bis (aryl) propane-type units.

On the other hand, the carbohydrates were depolymerized by acid hydrolysis during phase-separation treatment. It is extremely important to utilize the acid hydrolyzed carbohydrates because carbohydrates exist more than 70% in wood. Recently, the acid hydrolysis of lignocellulosics has become attractive again to solve energy issues and global warming caused by greenhouse gases, and to be applied to basic ingredient of biodegradable polymer and pharmaceutical supplies from hemicellulose.

The present paper deals with the separation behavior of lignocellulosics and the properties of the resulting carbohydrate fractions in the phase-separation system.

Experimental

Lignophenol

p-Cresol (10 ml/g wood) was added to spruce wood meals with stirring at room temperature. After 10 min, 72% sulfuric acid (15 ml/g wood) was added to the mixture and the vigorous stirring was continued at 30°C for 5-240 min. The reaction mixture was rapidly separated into the organic and aqueous phases by centrifugation. The organic phase was taken up and added dropwise to an excess amount of ethyl ether with vigorous stirring. The precipitates were dissolved in acetone and insoluble parts were removed by centrifugation. The acetone fraction was then concentrated under reduced pressure and added dropwise to an excess amount of ethyl ether with stirring. The precipitated lignin derivatives (lignocresol) were collected by centrifugation.

Carbohydrate composition

The aqueous phase after sulfuric acid hydrolysis was diluted to 3% sulfuric acid concentration with distilled water. The solution was boiled for 4 hours maintaining constant volume by frequent addition of water. After cooling, ribose was added as an internal standard. The solution was neutralized and desalted by barium hydroxide. After centrifugation, the supernatant was analyzed by the SHIMADZU High-Performance Liquid Chromatograph (HPLC) Reducing Sugar Analysis System under the following condition.

Column: Shim-pack ISA-07/S2504 (4mm ID. × 25 cm L.)
Eluent: A; 0.1 M potassium borate buffer (pH 8.0), B; 0.4 M potassium borate buffer (pH 9.0)
Flow rate: 0.6 ml/min
Gradient: A 100% to B 100% (2%/min)

Temp: 65°C
Post column reaction reagent: 1% *L*-arginine in 3% Boric acid at 0.5 ml/min
Reaction temp.: 150°C
Fluorescent detection: Ex-320 nm, Em-430 nm

The carbohydrate composition within the organic phase was analyzed as follows. The organic phase was added to ethyl ether with stirring. The precipitates were dissolved in acetone and insoluble parts were collected by centrifugation. The insoluble parts (ca 10 mg) were treated with 1 ml of 72% sulfuric acid at 30°C for 1 hour. The reaction mixture was diluted to 3% sulfuric acid concentration with distilled water and heated in an autoclave at 120°C for 1 hour. After cooling, ribose was added as an internal standard. After neutralization and desalting, the supernatant was analyzed by HPLC used for the aqueous phase.

Molecular weight distribution
The aqueous phase after sulfuric acid hydrolysis was neutralized and desalted by barium hydroxide. After centrifugation, the supernatant was analyzed by gel permeation chromatography (GPC) under the following condition.

Column: Asahipak GS-520HQ, GS-320HQ, GS-220HQ in a series (7.6 mm ID. × 30 cm L.)
Eluent: H_2O
Flow rate: 0.6 ml/min
Temp: 50°C
Detector: RI
Standard: Pullulan, maltose, maltotriose, maltopentaose, maltoheptaose

Results and Discussion

Carbohydrate composition
The wood fibers are composed of crystalline cellulose coated with noncrystalline hemicellulose, and hydrophobic lignin forming interpenetrating polymer network structure with carbohydrates. A part of lignin is linked to hemicellulose through ether or ester linkages. For complete hydrolysis, cellulose must be swollen by treatment with a concentrated acid such as sulfuric acid or phosphoric acid. However concentrated acid hydrolysis causes self-condensation of lignin, which limits dissolution of carbohydrate (the belt effect). Consequently, the treatment of 2-4 hours is nominated to dissolve cellulose selectively from wood cell wall.

Figure 2 shows the carbohydrate composition of the aqueous phase after the phase separation treatment. By the phase-separation treatment for 5 minute, the yield of neutral sugars from hemicellulose (mannose, arabinose, galactose and xylose) was about 34%. This yield was higher than total hemicellulose content in raw materials. However, the yield became lower with extended reaction time and the yield of neutral sugars from hemicellulose was about 19% after 60 min. This reason is that noncrystalline hemicellulose is preferentially hydrolyzed in the initial step of phase-separation treatment, followed by hydrolysis of crystalline cellulose with the increasing proportion of glucose.

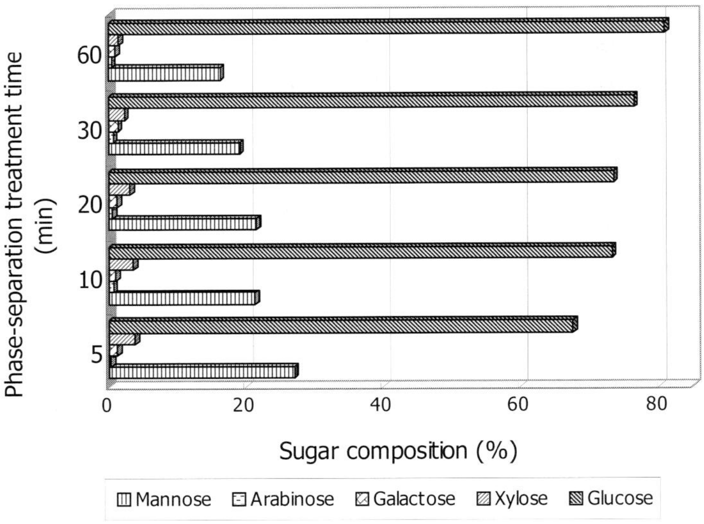

Figure 2 Carbohydrate compositions of the acid phase after the phase-separation treatment.

On the other hand, 11% of carbohydrates were in the cresolic phase after 5 minute phase-separation treatment although its yield decreased with longer treatment time (Figure 3). The yield of neutral sugars from hemicellulose was 40%. Because a part of lignin is linked to hemicellulose, the hemicellulose was separated into the cresolic phase with ligninophenol at an early stage of phase-separation treatment. However the hemicellulose content was decreased at the later stage because the linkages between lignin and hemicellulose were cleaved through the contact with acid at the interface between two phases.

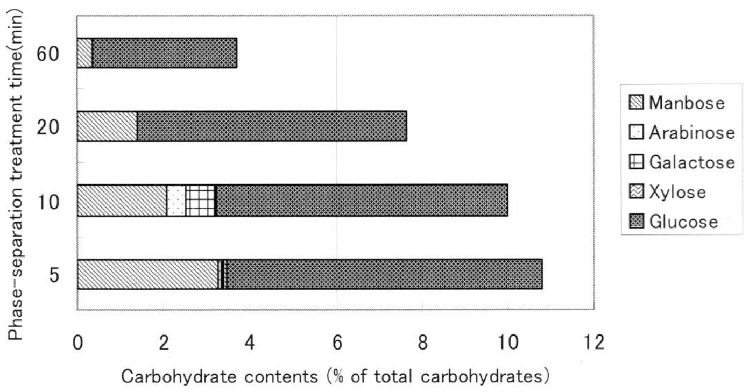

Figure 3 Carbohydrate compositions of the p-cresol phase after the phase-separation treatment.

Molecular weight distribution of carbohydrate

The molecular weight distribution of the carbohydrates in the sulfuric acid phase after phase-separation treatment was measured by GPC (Figure 4).

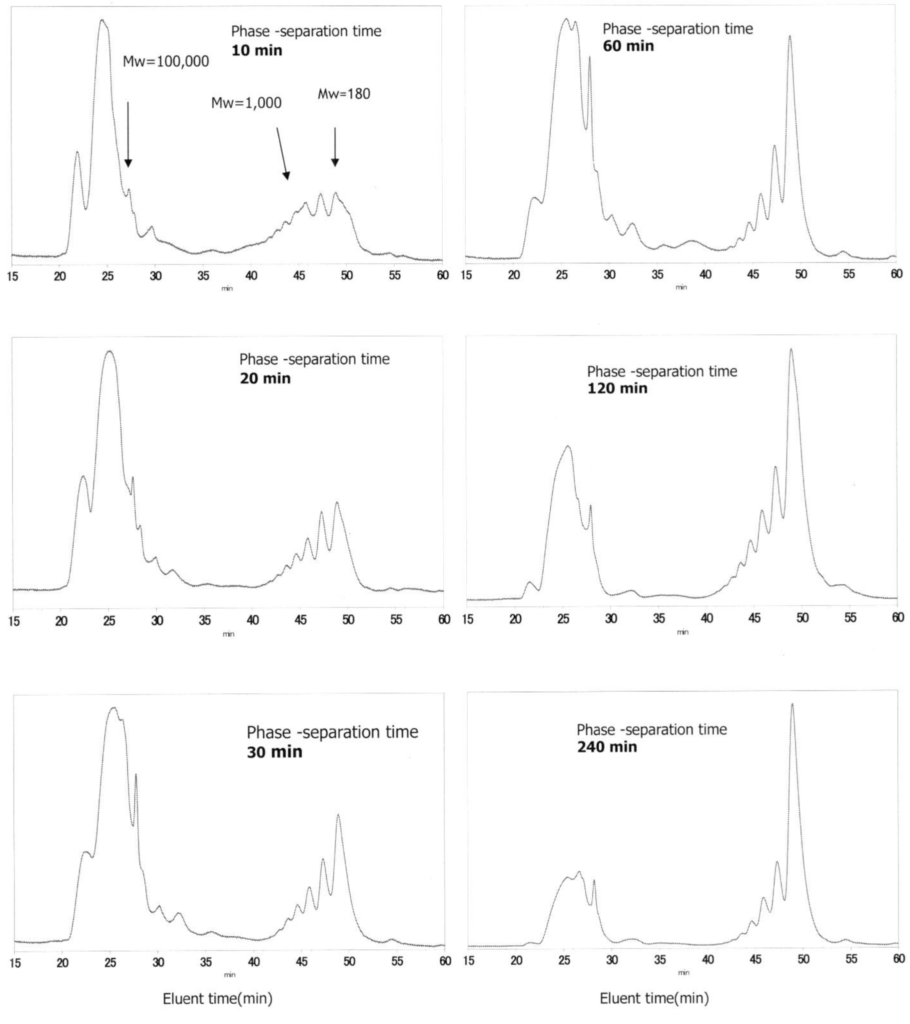

Figure 4 GPC profiles of carbohydrates in the acid phase.

The carbohydrates in the sulfuric acid phase were consisted of high molecular weight fractions (more than Mw=100,000) and low molecular fraction (less than Mw=2,000). The high molecular weight fractions were decreased with the phase-separation treatment time to be cleaved monomer and dimer fractions. The high molecular weight fractions were collected using GPC and the solution obtained was lyophilized. The resulting white powder was dissolved in D_2O and analyzed by ^{1}H-NMR. The ^{1}H-NMR spectrum of the high molecular weight fraction shows that this fraction was water-soluble glucose based polymer because the spectrum presented many broad signals for carbohydrate (3-5 ppm) (Figure 5). On the other hand, the low molecular weight fraction consisted of mixtures of the oligosaccharides with various molecular weights. A part of oligosaccharides was hydrolyzed to monosaccharides with extended reaction time (Figure 6).

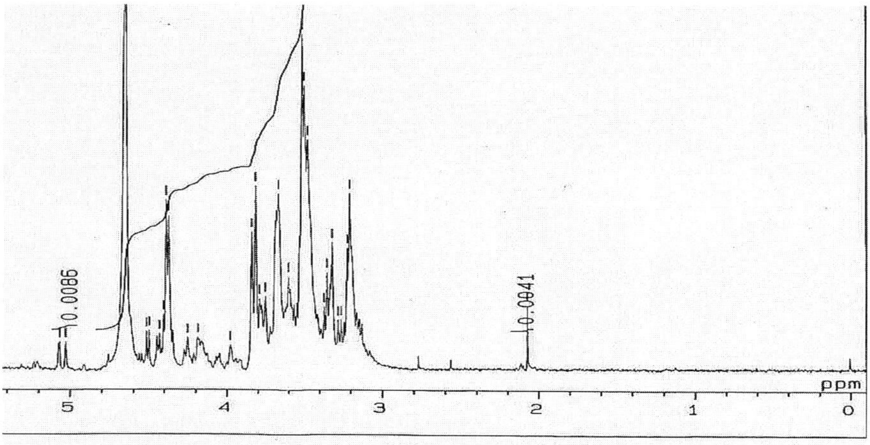

Figure 5 ^{1}H-NMR spectrum of high molecular weight fraction from the acid phase.

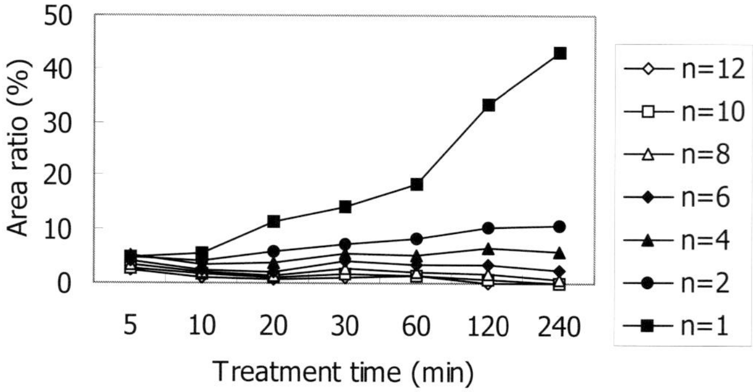

Figure 6 Polymerization degree of low Molecular Weight Fractions from the Acid Phase.

Conclusion

In the phase-separation reaction system composed of phenol derivatization and concentrated acid hydrolysis, lignin and carbohydrates were separated almost quantitatively from an interpenetrating polymer network within the cell wall into different phases within 30 minutes (Figure 7). In this system, lignin was solvated with phenol derivatives to be protected from the attack of concentrated acid.

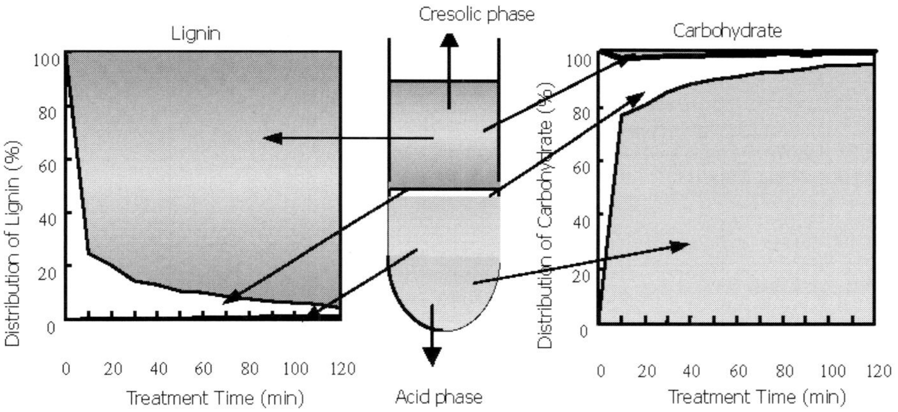

Figure 7 Material balance of lignin and carbohydrate in the phase-separation treatment.

Consequently, complicate deactivating modifications of lignin were greatly suppressed and lignin was phenolated selectively at a C-1 position. Through these modifications, the belt effect of lignin to restrict carbohydrates was released, so that carbohydrates were rapidly swollen and dissolved into concentrated acid. In this acid hydrolysis, a non-crystalline carbohydrate of hemicellulose was accessed and a part of hemicellulose was separated into the cresol phase with lignophenol at an early stage of phase-separation process. With extended reaction time, the carbohydrate content in the organic phase was reduced greatly because of the decomposition of lignin-carbohydrate complex into small fragments.

Separated carbohydrates were consisted of various sizes of oligosaccharides and water-soluble glucose based polymers with high molecular weight. The carbohydrate component can be converted readily into low molecular weight sugars by dilute acid hydrolysis. The monosaccharides can be also converted into ethanol through its fermentation. The ethanol is not only utilized as a fuel but also raw materials for chemical industries as ethylene through the dehydration process. The monosaccharides from hemicellulose are expected to be used as pharmaceuticals such as xylitol.

On the other hand, through this phase-separation system, native lignin is converted selectively and almost quantitatively to 1, 1-bis (aryl) propane-type polymers (lignophenol). The lignophenol has several unique functions discriminated from native lignin, which are highly phenolic property, no conjugated bonding, lighter color, thermoplasticity and high immobilization capability for proteins. Furthermore, this functionality of lignophenol can be secondarily controlled using a nucleophilic attack of C1-phenolic units to adjacent carbons, leading to the cascade-type utilization in various fields.

References

[1] Lipnsky, E. S. "Perspectives on Preparation of Cellulose for Hydrolysis" in "Hydrolysis of Cellulose: Mechanisms of Enzymatic and Acid Catalysis", eds. Jurasek, L. and Brown, Jr., R. D. (1979) 1.
[2] Johansson, B. and Miksche G. E. Acta Chem. Scand. 23 (1969) 924.
[3] Harkin, J. M. Advances in Chemistry Series 59 (1966) 65.
[4] Yasuda, S., Terashima, N. and Ito, T. Mokuzai Gakkaishi 26 (1980) 552.
[5] M. Funaoka and I. Abe Tappi Journal 72 (1989) 145.
[6] Funaoka, M. and Fukatsu, S., Holzforschung 50 (1996) 245.
[7] Funaoka, M. Polymer Int. 47 (1998) 277.

Chapter 5

Physical Aspects

An Essay on the Fine Structure of the Wood Cell Wall Related to the Physical Properties of the Recycled Paper

Hiroyuki Yamamoto[a], Yoichi Kojima[a], Takashi Okuyama[a], Toshihiro Ona[b] and Takayuki Okayama[c]

[a]*Graduate School of Bioagricultural Sciences, Nagoya University, Chikusa-ku, Nagoya 464-8601, Japan*

[b]*Graduate School of Bioresource and Bioenvironmental Sciences, Kyushu University, Hakozaki, Higashi-ku, Fukuoka 812-8581, Japan*

[c]*Graduate School of Agriculture, Tokyo University of Agriculture and Technology, Fuchu 183-8509, Japan*

Introduction

Remarakable progress in microscopic techniques from the 1950's to the 1960's enabled plant anatomists to clearly describe the fine structure of the wood cell wall. Nowadays, it is our common knowledge that each lamella can be approximated by a "two-phase structure", i.e. the reinforcing element of cellulose microfibril (CMF) and the matrix (MT) substance of lignin-hemicellulose compound. However, an exact image of the fine structure and the functions of each cell wall constituent has yet to be clarified. For better utilization of the wood as the engineering materials, it is indispensable that such characteristics should be revealed. However, it is still difficult to measure the complicate properties of each cell wall constituent directly as it is in the cell wall.

The authors' approach is to formulate the behaviors of the single wood fiber in its longitudinal direction based on the general theory of the micro-mechanics for the composite materials. Obtained formula contains several parameters which come from the higher-ordered structure of the cell wall and the microscopic properties of each constituent. When simulating the observed phenomena by using the obtained formula, we need to give concrete values to those parameters in a rational manner. Then, it is quite natural to consider that the values given in the simulation reflect intrinsic information on the fine structure and the internal properties of the cell wall constituents. From such a background, we have been analyzing macroscopic behaviors of the clear wood specimen especially for the mechanical properties in its axial direction.

Through those investigations, we obtained an interesting prediction on the interface structure between the CMF and the MT in the wood cell wall. This prediction was obtained from the moisture-dependencies of the longitudinal Young's modulus of the clear wood specimen [1, 2], which are expected to highly correlate with the recyclable properties of the pulp fiber [3, 4]. Thus, main topic in this report is a tentative image of the fine structure of the wood cell wall revealed by the simulation, and its role on the recyclable properties of the paper.

Mechanics for the deforming wood fiber

Models of the cell wall and wood fiber

A complex tube having three-layers, namely, compound middle lamella (CML), outermost layer (S1) and middle layer (S2) of the secondary wall, is used as a mechanically equivalent model of the wood fiber cell [2]. The CMF is oriented in a certain direction to the fiber axis in each layer of the secondary wall, however, it is randomly distributed in the CML, and we consider it to be mechanically isotropic.

Under magnification lesser than the ordinary scanning electron microscope, we can consider that the CMF are spatially dispersed uniformly in the cell wall lamella, and it forms a polysaccharide framework bundle with the oriented polyose. In a similar way, the lignin-hemicellulose compound forms the MT skeleton by being diffused in the cell wall lamella. Thus, it is hypothesized that both the CMF bundle and the MT skeleton occupy the same domain in the macroscopic limit. From the mechanical viewpoint, this idea can be formulated as follows [1, 2],

$$\sigma_{ij} = \sigma_{ij}^{f} + \sigma_{ij}^{m}, \qquad \varepsilon_{ij} = \varepsilon^{f}_{ij} = \varepsilon^{m}_{ij}, \qquad (1)$$

where σ_{ij}^{f}, σ_{ij}^{m}, and σ_{ij} are the stress tensors in the CMF bundle, the MT skeleton, and the cell wall layer as a whole, respectively. ε_{ij}, ε^{f}_{ij}, and ε^{m}_{ij} are their respective strains.

Basic equations

The anisotropic dimensional changes of the wood fiber model in the radial, the tangential, and the longitudinal directions are calculated under the assumption of the axisymmetrical (O-RTL) deformation [2]. In this paper, the dimensional change of the wood is caused by the increase in the strain energy due to an external load (an axial force) induction. Then, we assume the following constitutive equations where the CMF bundle and the MT skeleton in each layer must satisfy

$$\sigma_{ij}^{f} = C_{ijkl}^{f}\, \varepsilon_{ij}\,, \qquad \sigma_{ij}^{m} = C_{ijkl}^{m}\, \varepsilon_{ij}\,. \qquad (2)$$

In each layer, the eqs.(1) and (2) are combined, and the compatibility and equilibrium conditions for axisymmetry,

$$\varepsilon_{RR} - \varepsilon_{TT} = r\, d\varepsilon_{TT}/\, dr, \qquad \sigma_{RR} - \sigma_{TT} = -r\, d\sigma_{RR}/\, dr \qquad (3)$$

are applied. Then, fundamental formula is derived for arbitrary layer. Finally, the dimensional change of the wood fiber caused by the induction of the axial stress (P_L) can be solved as a following simple equation,

$$\varepsilon_L = f(\mathbf{p})\, P_L, \qquad \varepsilon_T = g(\mathbf{p})\, P_L. \qquad (4)$$

Thus, the longitudinal Young's modulus of the wood fiber model E_L and its Poisson's ratio ν_{LT} are calculated as follows:

$$E_L = P_L/\varepsilon_L = 1/f(\mathbf{p}), \qquad \nu_{LT} = -\,\varepsilon_T/\varepsilon_L = -\,g(\mathbf{p})/f(\mathbf{p})\,, \qquad (5)$$

where the quantity **p** is a parameter vector whose components are the layer thickness, the microfibril angle in the S2 layer (MFA), and the elastic moduli of the CMF bundle and the MT skeleton. For computing Eqs.(5), we need to optimize proper values and the moisture content (t) dependencies to those parameters so as to give a reasonable simulation. As mentioned in the Introduction section, we consider that the values and t-dependent patterns of the parameters derived in that way reflect intrinsic information on the fine structure and internal properties of the CMF and MT substance.

Result and Discussion

Moisture Dependency of the Elastic Modulus in wood

It is well known that the longitudinal Young's modulus of a clear wood specimen tends to decrease as the moisture content increases below the fiber saturation point (FSP), and becomes constant above the FSP. It is generally believed that the reduction of the Young's modulus is mainly due to the moisture dependent properties of the MT substance which consists of non-crystalline constituents.

Axial Young's modulus of the cellulose crystal (E_{cry}), isotropic Young's modulus of the MT substance (E_{matr}) and their contents in each layer are important factors in the simulation. Among them, only E_{matr} changes with the moisture content. According to Cousins [5, 6], E_{matr} takes a value of 2GPa under the sufficient moisture content, and monotonously increases up to 4 - 6GPa as the relative humidity decreases. On the other hand, elastic modulus of the cellulose crystal (E_{matr}) is assumed to be 134GPa, which is not affected by the moisture adsorption [7].

Under such an assumption, we calculated the moisture content dependency of E_L by using Eqs.(5). Results are shown in Figure 1(a). In order to obtain a simulation which is compatible to the experiment [8], we need to assume 24 - 28GPa as E_{matr} at the oven-dried state (ODS), which is pretty larger as compared to the standard value. If we accept both Cousins' experiments and the simplified cell model, we cannot help regarding this result as unreasonable crux. To explain this, two possibilities are pointed out [1, 2].

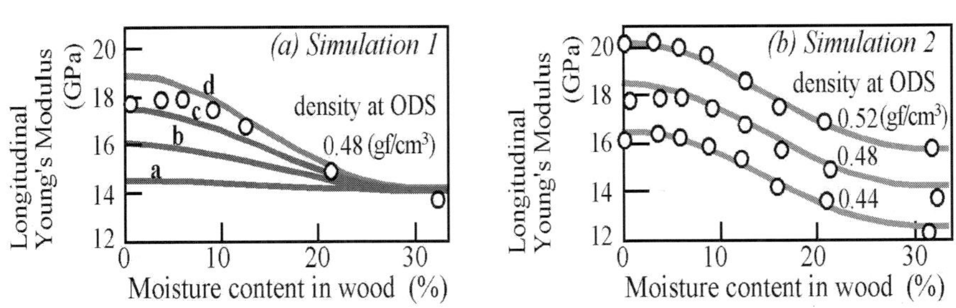

Figure 1 Moisture content dependencies of the longitudinal Young's modulus (E_L). Dots are experimental results from Picea sp. [8]. Curves are simulated results. MFA is assumed to be 10 deg. Simulation 1 was carried out under the possibility (a). In this case, E_{matr} increases from 2 GPa (at FSP) to a; 4, b; 12, c; 20, d; 28 GPa (at ODS). Simulation 2 was carried our under the possibility 2. E_{matr} increased from 2 GPa (at FSP) to 4 GPa (at ODS).

(a) Inside the actual cell wall, E_{matr} is 2GPa at FSP. However, at ODS, it is several times as rigid as the value from the isolated MT substance.
(b) Inside the actual cell wall, E_{matr} is 2GPa at FSP, and 4GPa at ODS. In addition, the CMF framework bundle becomes stiffer as the moisture content decreases.

If the possibility (a) could be correct, the reduction of the E_L due to water sorption should be more remarkable in specimen with large MFA than that with small MFA. However, we showed that this reduction becomes less distinct as MFA increases [9]. Thus, the possibility (a) is rather disputable.

As for possibility (b), this sounds a little strange since the CMF crystal would not react with water molecules. Then, to discuss this while we still accept 4GPa as the standard value of E_{matr} at ODS, the following hypothesis on the mechanical properties of the wood CMF is adopted [1, 2, 9]:

There is an unstable intermediate domain between stable crystal and stable amorphous domains in the wood CMF bundle, and such a domain fluctuates between quasi-crystal and quasi-amorphous states in accordance with the moisture adsorption. The quasi-crystal domain, which is mechanically close to the already-existing stable crystal domain, tends to increase with water desorption, then, Young's modulus of the CMF as a framework bundle tends to increase as the moisture content decreases.

Based upon this idea, we obtained a reasonable simulation as shown in Figure 1(b) when we assume that the content of the unstable domain in the secondary wall CMF reaches about one-third as much as the content of the stable crystal domain. It is imagined that deposition of the matrix substances, especially the cellulose-affinitive hemicellulose, induce the stable amorphous domain here and there in the surface of the CMF, and unstable domain is formed around the stable amorphous domain in the wood CMF. Based on such discussions, we show an imaginative model of the fine structure of the cell wood wall in Figure 2.

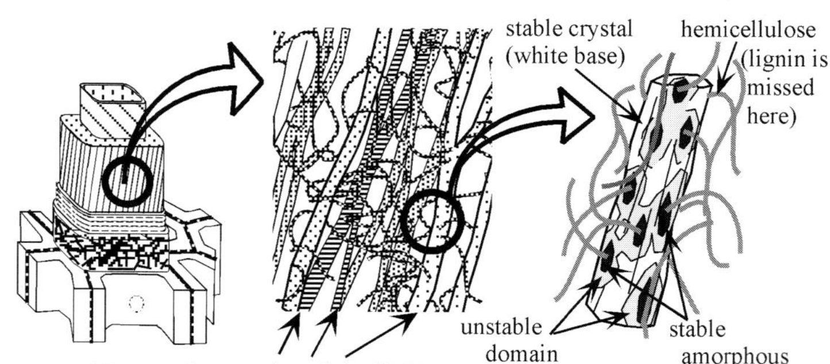

Figure 2 An imaginative model of the fine structure of the wood secondary wall.

Recyclable properties of the paper and the fine structure of the wood cell wall

The fibers in the dried paper sheet are glued each other by the force of hydrogen bond. However, once the dry paper is immersed in the water, the cell wall swells again, then, the paper is disjointed into fibers in the water. After several repetitions of the drying and wetting process, cornification gradually progresses in the fiber wall. As the result, the swelling ability of the fiber wall decreases with the number of recycle use, at the same time, the ability of the hydrogen bond in the fiber wall remarkably decreases. Thus, the tensile index of the paper sheet tends to decrease with the number of recycle use.

Okayama et al. [4] found that the tensile index of the paper manufactured from *Acacia spp.* rapidly decreases with the repetitions of the recycle treatment, however, the zerospan tensile index of the paper, which is an indicator of the strength of individual fiber, does not decrease with the repetitions of the recycle treatment. From this result, they considered that a decrease in the strength of the recycled paper is attributable not to the decrease in the tensile strength of the fiber but to the decrease in the coherence between the fibers in the sheet, and they supposed that such a decrease in the cohesion between fibers is induced by the cornification of the fiber wall.

Ona et al. [3] reported that the recyclable index of the papers manufactured from 9.5-year-old plantation trees of *Eucalyptus globulus* are scattered unevenly in spite that the strengths of the virgin papers become uniform. Moreover, they found that the recyclable index of the paper showed a high positive correlation with the percentage of hemicellulose, on the other hand, a high negative correlation with the cellulose percentage in the raw material. Those results suggest that the hemicellulose in the fiber wall plays a certain role on inhibiting the progress of the cornification in the fiber wall, that is to say, further crystallization of the cellulose microfibril due to the repetitions of the recycle treatment.

It is considered that their discoveries are highly related to our prediction on the fine structure of the wood cell wall. It is expected that an irreversible crystallization occurs in the unstable domain of the CMF with the repetition of the drying and wetting process, which is a major cause of the cornification of the fiber wall in the recycled paper.

References

[1] Yamamoto, H. Cellulose Commun. 7 (2000) 9.
[2] Yamamoto, H. and Kojima Y. Wood Sci. Technol. 36 (2002) 55.
[3] Ona, T., Kojima, Y., Kawana, J., Kibatani, Y., Ishikura, Y. and Okayama, T. *in this book.*
[4] Okayama, T., Kibatani, Y., Nakayama, K., Kojima, Y. and Ona, T. *in this book.*
[5] Cousins, J. Wood Sci. Technol. 10 (1976) 9.
[6] Cousins, J. Wood Sci. Technol. 12 (1978) 161.
[7] Nishio, T., Takano, K. and Nakamae, K. J. Polym. Sci. B. Polym. Physic. 33 (1995) 1647.
[8] Kollmann, F. and Krech, H. Holz Roh. Werk. 18 (1960) 41.
[9] Kojima, Y. and Yamamoto, H. Wood Sci. Technol. *in press.*

Wood Qualities of Some Tropical Plantation Species - Does Hastening the Growth Debase the Quality?

Takashi Okuyama[a], Hiroyuki Yamamoto[a], Imam Wahyudi[b] and Yusuf Sudo Hadi[b]

[a]Graduate School of Bioagricultural Sciences, Nagoya University, Nagoya 464-8601, Japan
[b]Faculty of Forestry, Bogor Agricultural University, Bogor, Indonesia

Introduction

It is the most important and an urgent task for humankind to establish the consistency between maintenance of global environment and sustainable supply of forest resources. In order to accomplish this task it is very important to encourage establishing forest plantations especially in the tropical regions. Recently, plantation forests have become very common in many tropical countries [1]. These plantations mainly aim to produce short rotation raw materials for pulp and paper. However to make this effort successful, it is necessary to establish an economical cycle between producer and consumer sectors by raising the added value of the plantation species. Developing timber use of the species is an essential way to raise the economical value of the short rotation trees. The growth rate of the tropical plantation species is faster than that of Temperate Zone species. Therefore, it is worried about that the inferior qualities due to the larger growth rate will take place during lumber processing. This investigation is focused mainly to the growth stress because the growth stress is the most important obstacle facing in the first place at sawing. Better understanding of the growth stress will contribute to raising timber value by linking the genetic improvement of tree, better afforestation technology and improved wood processing technologies [2]. From the above background, we have been conducting investigation on the effect of extremely high growth rate upon the growth stress and other properties of plantation species in Indonesia.

Results and Discussion

It is very important for forester to know whether the growth rate affects the wood qualities or not. If wood qualities become inferior due to silvicultural treatments that accelerate tree growth e.g., fertilizing, it will be disadvantage to them. The relationships between density and the averaged growth rate of each species are shown in Figure 1. The growth rate taken as the quotient of dbh over tree age found on the x-axis, reflects the average growth rate of several trees. All species seem to have tendency to grow faster at younger stage of growth. For instance in acacia mangium (*Acacia mangium* Willd.), the average growth rate of 4-year-old trees was 2-5 cm/year, while for 8-10-year-old trees, 1.2-2.6 cm/year increment was observed. The same tendency was also observed in young teak (*Tectona grandis* L.). In the case of falcata (*Paraserianthes falcataria* (L.) Nielsen), the trend was not obvious because the age difference among trees was very small. Nevertheless, 3-year-old trees show faster growth rate than the 6-year-old trees. Figure 1 indicates that density is not influenced by the growth rate.

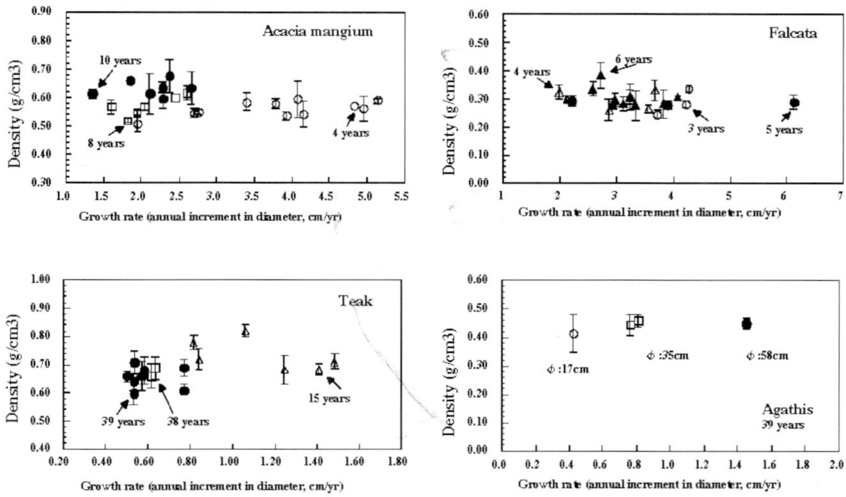

Figure 1 Relationships between density at periphery and average growth rate of individual trees.

Radial distributions of density for various growth rates are indicated in Figure 2. Acacia mangium for example, when 4-year-old trees were compared to 10-year-old trees of the same diameter, it was observed that density increases from pith to bark. Despite the age differences between the two samples, similar distribution patterns were observed. Density of 10-year-old trees was higher than 4-year-old trees, nevertheless both follow the same trend. At the outer portion of the stem, the older trees showed a flattish distribution while younger trees a steep distribution was observed. Thus, density distribution alone is not a good determinant of the maturation age of the trees because of the flat curve 2/3 of the diameter of the stem. This was also observed in falcata. However their microfibril angle shows steep distributions, then 10 year-old trees cannot be considered yet as mature. In case of teak, 10-15 cm from the pith or at the 13th and 15th growth ring, the density was lower but showed flattish distribution. At the same position, microfibril angle distribution was more or less uniform. Therefore in teak, it was considered that the maturity of the stem starts on the 13th-15th year. As for agathis (*Agathis borneensis* Warb.), despite the existence of compression wood, the density distribution was almost constant from the pith to the bark regardless of the variation in the diameter and the growth rates among experimental trees. There was a clear relation between the amount of released strain and the MFA, as shown in the past research [3]. As the absolute value of the released strain of acacia mangium increases, MFA decreases (below 8 degrees). The same tendencies were observed also in the cases of both falcata and teak. As for agathis, expansive strains were observed at large MFA region where the compression wood exists. This result confirms well our hypothesis on the generation mechanism of growth stress. Growth stress in the cell wall is generated as a compound stress generating from the tensile stress created in the microfibrils along its axial direction and an isotropic compressive stress generated in the cell wall matrix between microfibrils [3-5].

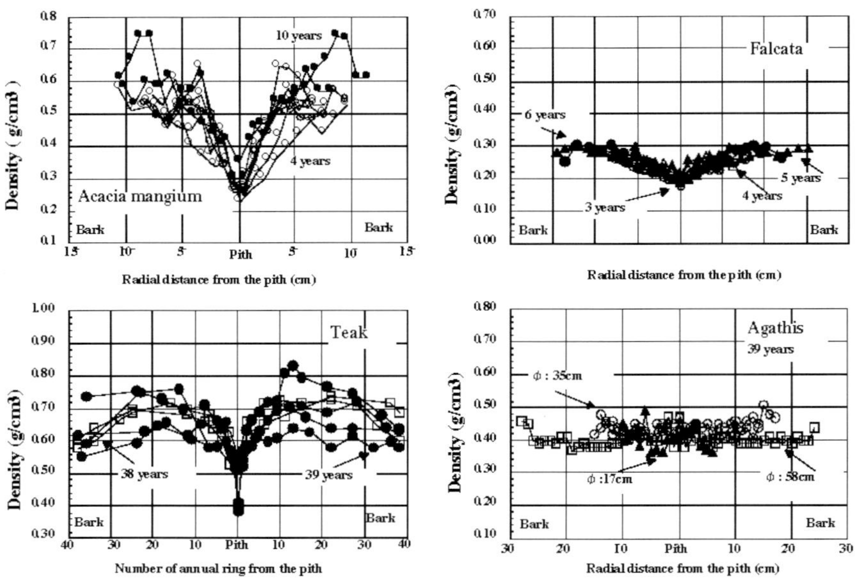

Figure 2 Radial distributions of density.

Figure 3 shows the released strains after eliminating the data that have reaction wood decided by values of MFA. These include MFA values below 8 degree in acacia mangium and 10 degree in falcata because of the presence of gelatinous fibers. For teak, gelatinous fiber was not recognized but the region showed high released strain coincides to the low microfibril angle region as shown in *Magnoliaceae* [3]. Agathis showed positive values in the region of compression wood. All species, the average released strain values were low and more or less uniform regardless of the growth rate. Then we can conclude that the growth stress is not affected by the growth rate. Accordingly, avoiding the formation of tension wood is an effective way to reduce the growth stress generation.

Moreover, accelerating tree growth causes neither high growth stress nor low density.

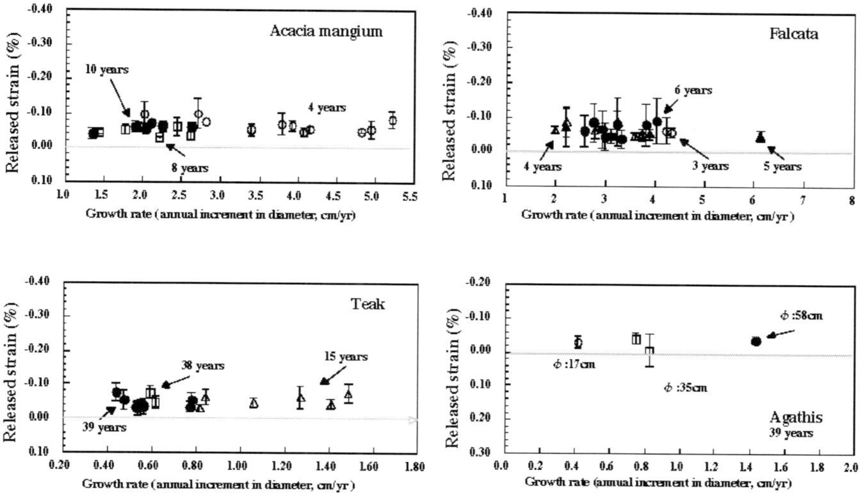

Figure 3   Released strains after eliminating the data in reaction wood.

References

[1] FAO, Global Forest Resources Assessment 2000, Main report: FAO Forestry paper (2000) 140.

[2] Okuyama, T., Kanagawa, Y. and Hattori, Y. Mokuzai Gakkaishi 33 (1987) 837.

[3] Okuyama, T., Yamamoto, H., Yoshida, M., Hattori, Y. and Archer, R.R. Ann. Sci. For. 51 (1994) 291.

[4] Yamamoto, H., Okuyama, T. and Yoshida, M. Mokuzai Gakkaishi 39 (1993) 375.

[5] Yamamoto H., Okuyama, T. and Yoshida, M. Mokuzai Gakkaishi 39 (1993) 118.

Research on Efficient Use of Forest Products - Engineering Evaluation of Good Tactile Warmth for Wood

Yoshihiro Obata[a], Kazutoshi Takeuchi[b], Hideaki Sugino[c] and Kozo Kanayama[a]

[a]*Advanced Wood-based Material Technology Group, Institute for Structural and Engineering Materials, National Institute of Advanced Industrial Science and Technology, 2266-98, Anagahora, Shimoshidami, Moriyama-ku, Nagoya 468-8560, Japan*

[b]*Temporary Research Staff dispatched to* [a]

[c]*NEDO Industrial Technology Researcher in* [a]

Introduction

Recently, the concept of "sustainable development" has been taken seriously in the fields of resources, energy and environment, which are commodities essential for continuous development of human society. The sustainable use of resources without damages to environment is our challenge and duty to meet not only the need of the present society but also one of the future generations. From viewpoint of this idea, the importance of forest and forest products has been reevaluated. Because wood as one of the forest products is a sustainable industrial resource if the sustainable forestry, *i.e.*, the cycle of felling, planting and growing trees would be continued, against the limited mineral resources such as metals, oils and gasses, which are estimated to be mined all up in the near future [1]. Wood is also the sustainable energy source. In fact, fifty-five percent of sawn wood production in the world was used as fire wood at mainly the developing regions in the year 1997 [2]. In addition to the problem of mineral resource shortage, we also face to the serious global environment problem, *i.e.*, the global warming, caused by greenhouse effect with several gasses such as carbon dioxide, chlorofluorocarbons, methane, nitrous oxide and so on. Especially, carbon dioxide is estimated to have 57% greenhouse contribution [3]. Forest can fix carbon from carbon dioxide by photosynthesis against this environmental problem. So forest supplies us forest products as industrial resource and energy source without damage to the environment. But the role of forest against the global warming and the usage of wood as resource seem to conflict each other. Generally speaking, younger and sound forests have the higher ability to fix carbon from carbon dioxide [4]. The stage from felling into planting trees in the cycle for sustainable forestry means to replace the older trees with the younger ones. The affirmative usage of wood as the substitute of mineral resources can save the mineral resources and contribute to continue the sustainable forestry, *i.e.*, the cycle of felling, planting and growing trees. Then we can expect the sustainability of forest and forest products. For the realization of this philosophy, the technologies to improve the bad points of wood and to evaluate of good points of wood have been expected.

Now we have two serious problems on wood in Japan. One problem is the increase of insufficient- and no-thinned forests because of the lower commercial value of thinned wood. This situation can give us the worst scenario that we will not have the sound forests in Japan in the near future. The technology to give the commercial value to the thinned wood is expected to carry our the thinning of forests. Another problem is the wooden building waste. The amount of it was estimated to be 5 million ton in the year

2000 [5]. The incineration of wooden waste has been regulated, because of the dioxin emission. The recycle use of wooden waste is expected strongly. Compressed wood and mean density fiberboard (MDF) are prospective products for the effective use of thinned wood and the recycle use of wooden waste. These improved or recycled wooden products are desirable to have added values of good points as wood such as good tactile warmth, because one of their applications is flooring panel.

In this paper, we treated the evaluation of good tactile warmth of wood as one of many good points of wood. The aim of this study is to obtain the relation between the tactile warmth and the material properties. There have been some reports on the relationship between the tactile warmth and the physical quantities [6-9]. Their measure of tactile warmth is based on statistics of the judge of experimental subjects. Some of their papers have paid their attention to the relation between the sensory tactile warmth and the thermal conductivity. They pointed out that materials with smaller thermal conductivity were felt warmer. Okajima et al. treated building materials containing wood and metals but they concluded that it was difficult to find out the simple expression to show the relation between the thermal conductivity and the sensory tactile warmth [7]. Harada et al. reported that the sensory tactile warmth of wood has a high and negative linear-correlation with the logarithm of thermal conductivity [9]. Harada's result has been generally accepted as a relation between the sensory tactile warmth and the physical quantities of wood [10]. But it is impossible to compare the tactile warmth of wood and metals from their results, because their measure of sensory tactile warmth is available in the closed group of materials that the sensory inspection was done.

We have tried to find out the relation between the tactile warmth and the material properties from the theoretical analysis of the heat transfer phenomenon common to wood and other materials, and to establish the evaluation system of tactile warmth of materials [11]. The measure of tactile warmth is desirable to satisfy the following conditions: (1) it can be determined absolutely from physical quantities, (2) it can be available for not only woods but also other materials and (3) it is easy to understand in order to explain the good point of wood to the end users for more use of wood as substitute of mineral resources. Firstly, we show the analysis of governing heat transfer phenomenon on tactile warmth. We derive the relationship between the contact surface temperature and the thermal effusivity as the measures of tactile warmth. Then we review the sensory tactile warmth of wood in Ref. [9] with our measures. The other experimental knowledge on tactile warmth is also discussed with these properties. We also discuss the cases not only humans hand but also humans sole touches with wood to assume that compressed thinned wood and recycled MDF are used for flooring panel.

Theory

Let us consider the governing heat transfer phenomenon when our hand comes in contact with various materials. We put back our hand unconsciously and quickly when we get in touch with something too hot or too cold. It suggests that the tactile warmth is sensed sharply immediately after the contact and then it is judged whether the material is safety or not to keep our hand in contact with it. On the other hand, the human's sensory organs to sense warmth and coldness are located at 0.4mm and 0.2mm below the human's skin, respectively [12]. So the governing heat transfer phenomenon on tactile warmth is the phenomenon, which occurs near the contact surface in a short time after the contact. Then we can derive an analytical model for such transient response approximately with the transient one-dimensional heat conduction problem when two semi-infinite bodies, which have different material properties and different uniform initial temperatures, are placed in contact at their free surfaces each other.

Now we consider the heat transfer problem that two semi-infinite bodies come in contact with each other at 0 in x-coordinate and the temperatures become same at the surfaces of both semi-infinite bodies. Then the basic equations of transient one-dimensional heat conduction, the initial conditions and the boundary conditions are given as follows,

Basic equations:

$$\lambda_{\mathrm{H}} \frac{\partial^2 T_{\mathrm{H}}(t,x)}{\partial x^2} = C_{\mathrm{H}} \rho_{\mathrm{H}} \frac{\partial T_{\mathrm{H}}(t,x)}{\partial t} \tag{1}$$

$$\lambda_{\mathrm{M}} \frac{\partial^2 T_{\mathrm{M}}(t,x)}{\partial x^2} = C_{\mathrm{M}} \rho_{\mathrm{M}} \frac{\partial T_{\mathrm{M}}(t,x)}{\partial t} \tag{2}$$

Initial conditions:

$$T_{\mathrm{H}}(0,x) = T_{ini\mathrm{H}} \ \text{ at } \ t=0 \tag{3}$$

$$T_{\mathrm{M}}(0,x) = T_{ini\mathrm{M}} \ \text{ at } \ t=0 \tag{4}$$

Boundary conditions:

$$T_{\mathrm{H}}(t,0) = T_{\mathrm{M}}(t,0) = T_{cs}(t) \ \text{ at } \ x=0 \tag{5}$$

$$[\lambda_{\mathrm{H}} \frac{\partial T_{\mathrm{H}}(t,x)}{\partial x}]_{x=0} = [\lambda_{\mathrm{M}} \frac{\partial T_{\mathrm{M}}(t,x)}{\partial x}]_{x=0} \ \text{ at } \ x=0 \tag{6}$$

where T is the temperature, x is the coordinate, t is the time, λ is the thermal conductivity, C is the specific heat, ρ is the density, T_{ini} is the initial temperature before contact and T_{cs} is the contact surface temperature. The subscripts of H and M denote human beings and material, respectively.

The procedure of analysis is well known and the contact surface temperature T_{cs} is given with a very simple expression as follows [13];

$$T_{cs} - T_{ini\mathrm{M}} = \frac{T_{ini\mathrm{H}} - T_{ini\mathrm{M}}}{1 + \eta_{\mathrm{M}} / \eta_{\mathrm{H}}} \tag{7}$$

where,

$$\eta = \sqrt{\lambda C \rho} \tag{8}$$

which is the thermal effusivity. We should note that T_{cs} in Eq. (7) is constant on time. The heat flux from human's body to material is also given as follows;

$$q(t) = \frac{\eta_{\mathrm{H}}\eta_{\mathrm{M}}}{\eta_{\mathrm{H}} + \eta_{\mathrm{M}}} \frac{T_{ini\mathrm{H}} - T_{ini\mathrm{M}}}{\sqrt{\pi t}} = \eta_{\mathrm{M}} \frac{T_{\mathrm{cs}} - T_{ini\mathrm{M}}}{\sqrt{\pi t}} \tag{9}$$

Numerical result and discussion

Numerical conditions and material properties

Let us consider the situation that a human's hand comes in touch with the various materials at room temperature. Considering that a man feels moderate warmth when his skin's average temperature is 33 – 34 °C [14], we give the following initial temperatures to hand and materials:

$$T_{ini\mathrm{H}} = 32\ ^\circ\mathrm{C},\quad T_{ini\mathrm{M}} = T_{room} = 20\ ^\circ\mathrm{C}$$

$T_{ini\mathrm{H}}$ is 1-2 degrees centigrade lower than the average skin's temperature since the hand is an end organ in a human body. The material properties of material such as wood, metals and so on are incited from references [9, 15]. Thermal effusivity of human body is different for each part of it. We use thermal effusivity of palm as η_{H} with [16]

$$\eta_{\mathrm{palm}} = 1.263\ \mathrm{kJ/(m^2 \cdot s^{1/2} \cdot K)}$$

Evaluation of the sensory tactile warmth on woods with contact surface temperature

The sensory tactile warmth on wood was reported by Harada et al. in detail [9]. They showed that the sensory tactile warmth of woods is proportion to the logarithm of the wood's thermal conductivity. We review their result with the contact surface temperature. Figure 1 shows the relationship between their sensory tactile warmth and our contact surface temperature. The sensory tactile warmth has a high positive correlation with the logarithm of the temperature difference between the contact surface temperature and the material's initial one. The relationship is expressed as follows:

$$S \propto K_1 \log(T_{\mathrm{cs}} - T_{ini\mathrm{M}}) \propto K_2 \log(1 + \eta_{\mathrm{M}} / \eta_{\mathrm{H}})^{-1} \tag{10}$$

where K_1 and K_2 are constants. Eq. (10) is coincident to the Fechner's formula, which shows a relation between state of mind S and stimulus R as follows:

$$S = K \log R \quad (K : \mathrm{constant}) \tag{11}$$

It suggests that the contact surface temperature and the thermal effusivity can be used as measures of the tactile warmth.

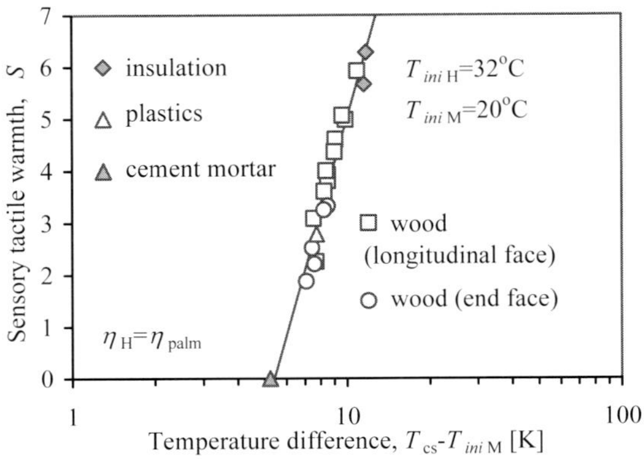

Figure 1 Sensory tactile warmth and contact surface temperature.

The thermal conductivity in fibrous direction of wood is 2.25-2.75 times of the vertical one to it [17]. Harada et al. treated the influence of the wood's anisotropic thermal conductivity upon the sensory warmth and reported that the touch with end face of wood felt colder than the touch with longitudinal face of the same wood. But Table 1 shows that the order of thermal conductivity does not correspond to the order of the sensory tactile warmth in a mixed contact system with end and longitudinal faces for some woods. On the other hand, the order of thermal effusivity coincides to the order of the sensory tactile warmth. This result suggests that the thermal effusivity evaluate tactile warmth of wood more accurately than the thermal conductivity.

Table 1 Sensory tactile warmth and material properties.

Materials	Face	S	λ [W/(m·K)]	η [kJ/(m^2·s$^{1/2}$·K)]
Seraya wood	long.	4.36	0.190	0.416
Seraya wood	end	3.25	0.350	0.571
oak wood	long.	3.09	0.330	0.742
oak wood	end	1.88	0.486	0.874

Legend: S; Sensory tactile warmth, λ; Thermal conductivity, η; Thermal effusivity

Figure 2 shows the relationship between the thermal effusivity and thermal conductivity for the material group of insulations, woods and others such as glasses, rocks, metals and alloys. Their relationships between both material properties are given with approximated lines. Although their relations are different for each material gropes, one for wood is given as follows.

$$\eta_M = C\lambda_M \quad (C:\text{constant}) \qquad (12)$$

Then Eq. (10) gives us the following result:

$$S \propto -\log(\lambda_M + C_1) \quad (C_1 : \text{constant}) \tag{13}$$

Equation (13) means that the sensory warmth for wood is in proportion to the logarithm of the linear function of thermal conductivity. It doesn't conflict with the result by Harada et al. on the tactile warmth of wood. But we should use Eqs. (7) and (9) based on thermal effusivity when we compare the tactile warmth of wood with other materials.

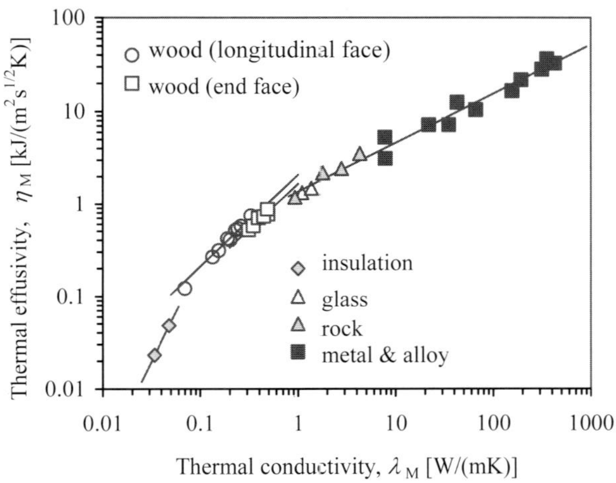

Figure 2 Thermal effusivity and thermal conductivity.

Contact surface temperature for woods and other materials

Eq. (7) is applicable to not only wood but also other materials. Figure 3 shows the relationship between the contact surface temperature and the thermal effusivity for various materials. The solid line represents the contact surface temperature when the thermal effusivity of palm is used as η_H in Eq. (7). The line shows that materials with smaller thermal effusivity are felt warmer than that with larger one. It shows also that the difference of contact surface temperature between different woods is very large. The result explains our experience that we can distinguish the difference of tactile warmth easily for different woods in spite of the little difference of their thermophysical properties. Copper was used as a heat source instead of human's palm in some experimental works [7, 9]. The dotted line shows the contact surface temperature in the case that the thermal effusivity of copper is used as η_H instead of palm's one. It is difficult to distinguish the contact surface temperatures for different woods in this case. Another narrow dotted line shows the contact surface temperature in the case that the thermal effusivity of sole is used as η_H. The thermal effusivity of sole is smaller than that of palm. The contact surface temperature of sole is lower than that of palm.

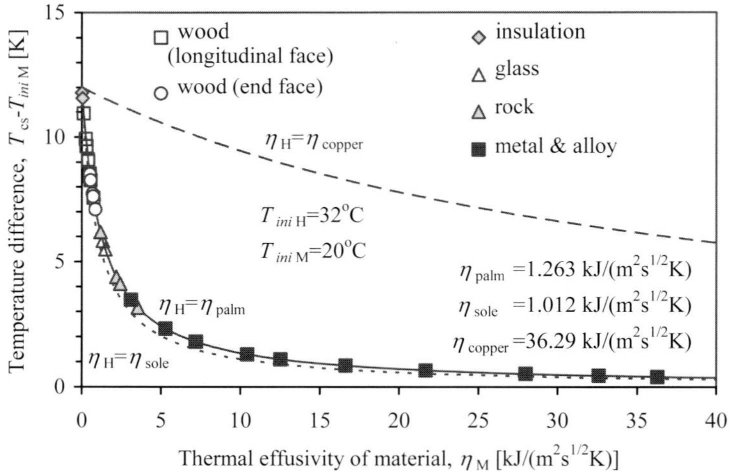

Figure 3 Contact surface temperature for various materials.

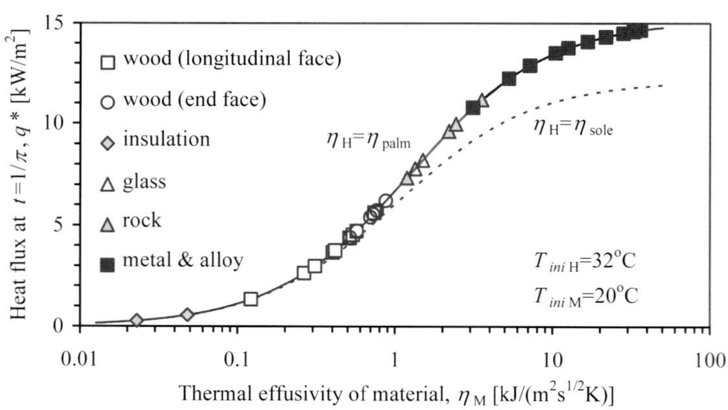

Figure 4 Heat flux from human body to various materials.

Figure 4 shows the heat flux from hand to materials given by Eq. (9). The heat flux at $t=1/\pi$ is shown in the figure, because the heat flux is a function of time. The horizontal axis is in logarithmic scale. The slope of the curves is very gradual for the material groups of not only metals but also insulations. This means that metals and insulations are difficult materials to distinguish each other in the same material group by tactile warmth. On the other hand, the slope for wood is very sharp to distinguish each wood easily by tactile warmth. It implies that wood be sensible and comfortable material for human beings.

Figure 5 shows the contact surface temperature with some materials for the various thermal effusivities of human body. There are differences among the thermal effusivities of palm, sole and forehead. It implies that there must be thermal effusivity among individuals. The figure shows that the smaller η_H feels woods colder than the larger one. But the smaller η_H can choose the warmer wood more easily than the larger one, because the difference of contact surface temperature between two woods is larger for the smaller η_H than the larger one.

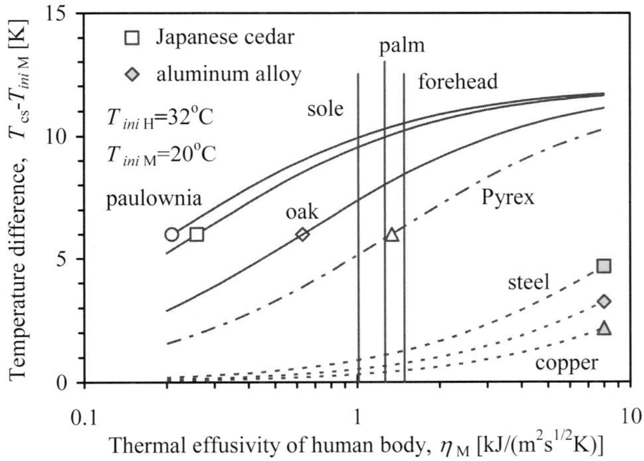

Figure 5 Influence of different thermal effusivity of human body.

Figure 6 shows the contact surface temperature when humans hand comes in contact with several materials at the various initial temperatures. The right side term in Eq. (7) is almost 0 for metals because of their large thermal effusivities. So the contact surface temperature is same to the initial temperature of material. This is the reason that metals are felt too cold in winter and too hot in summer to touch. On the other hand, wood has moderate tactile warmth regardless of the seasons. The contact surface temperature with wood heated at 100°C is not so high. This result explains our experimental knowledge that we can treat the heated wood easily with bare hand.

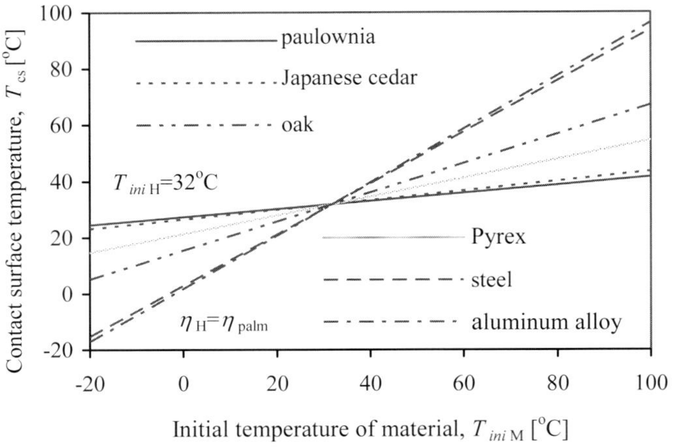

Figure 6 Contact surface temperature to materials with high/low initial temperatures.

Conclusion

We introduced that wood is a prospective material against the shortage of resources and the global warming from the viewpoint of sustainable development. We also made mention that the effective use of thinned wood and the recycle use of wooden building waste are serious problem. We treated tactile warmth of wood in order to encourage the wood in use more widely as substitute of other industrial materials and to give the improved or recycled wood the added value. We give the concluding remarks as following from the discussion on the evaluation of tactile warmth of wood and other materials.

(1) The contact surface temperature decreases for the materials with higher thermal effusivity.
(2) The sensory warmth of wood is proportional to the logarithm of the temperature difference of contact surface temperature from initial temperature of material.
(3) The thermal effusivity evaluates sensory warmth properly in a mixed contact system of end and longitudinal faces of wood.
(4) The relationship of the contact surface temperature and the thermal effusivity explains rationally that woods are felt much warmer than metals. It also explains that each wood has large difference of tactile warmth from other species of wood in spite of the small difference of their material properties.
(5) The relationship of the contact surface temperature and the thermal effusivity explains rationally that wood has good tactile warmth in spite of seasons, although metals are felt too hot in summer and too cold in winter to touch.

As a result of the above remarks, we propose the contact surface temperature and the thermal effusivity as the engineering measures to evaluate the tactile warmth of wood and other materials objectively. We expect that this result can be useful to control the and give the good tactile warmth to compressed wood and MDF.

References

[1] Kanayama, K. Wood Industry 52 (1997) 446.
[2] Forestry Agency, Annual Report on Trends of Forestry in Fiscal Year 1999 (in Japanese), Association of Agriculture & Forestry Statistics, Tokyo (2000) 137.
[3] Masters, G. M. "Introduction to environmental engineering and science", Prentice Hall, Englewood Cliffs (1991) 389.
[4] Sugiyama, M. Wood Industry 54 (1999) 440.
[5] Nakajima, S. Wood Industry 57 (2002) 186.
[6] Wada, Y., Oyama, T. and Imai, S. "Psychology handbook on sensation and perception (in Japanese)", Seishin-shobo, Tokyo (1969) 15-16, 778, 807.
[7] Okajima, T., Tanahashi. I., Yasuda, T. and Takeda, Y. Trans. Architec. Inst. Japan 245 (1976) 1.
[8] Matsui, I. and Kasai, Y. Trans. Architec. Inst. Japan 263 (1978) 21.
[9] Harada, Y., Nakado, K. and Sadoh, T. Mokuzai Gakkaishi 29 (1983) 205.
[10] Imamura, Y., Kawai, S., Norimoto, M. and Hirai, T. "Wood and woody materials (in Japanese)", Toyo-shoten, Tokyo (1997) 301-303.
[11] Obata, Y., Kohara, M., Furuta, Y., Kanayama, K. Mokuzai Gakkaishi 46 (2000), 137-143.
[12] Japan Society of Mechanical Engineers. "Biomechanics (in Japanese)", Ohmsha, Tokyo, (1991) 213-218.
[13] Frank, P. I. and David, P. D. W. "Fundamentals of heat and mass transfer, 3rd ed.", John Wiley & Sons, New York (1990) 259-262.
[14] Bioengineering Publishing Committee. "Bio- engineering (in Japanese)", Baihukan, Tokyo (1992) 54.
[15] Japan Society of Thermophysical Properties. "Thermophsical property handbook (in Japanese)", Yokendo, Tokyo (1990) 22-27, 64, 450-451, 493.
[16] Yokoyama, S. "Heat Transfer Phenomenon in Living Body (in Japanese)", Hokkaido Univ. Press, Sapporo (1993) 75.
[17] Nakado, K. "Wood Engineering (in Japanese)", Yokendo, Tokyo (1985) 113-116.

Chapter 6

Biochemical Aspects

Possible Effects of Properties in Polyphenol Oxidases on Rooting Ability of *Eucalyptus camaldulensis* Cutting Shoots

Shinso Yokota[a], Kohei Nakayama[a,b], Akiko Sagawa[a], Fumiko Urabe[a], Toshihiro Ona[c], Takayuki Asada[d] and Nobuo Yoshizawa[c]

[a]*Faculty of Agriculture, Utsunomiya University, 350 Mine-machi, Utsunomiya, 321-8505, Japan*
[b]*Present address: Faculty of Agriculture, Tokyo University of Agriculture and Technology, 3-5-8 Saiwai-cho, Fuchu 183-8509, Japan*
[c]*Graduate School of Agriculture, Kyushu University, 6-10-1 Hakozaki, Higashi-ku, Fukuoka 812-8581, Japan*
[d]*Kameyama Research Center, Forestry Research Institute, Oji Paper, 24-9 Nobono-cho, Kameyama 519-0212, Japan*

Introduction

The cuttings from some old *Eucalyptus* trees have inferior ability to root, leading to difficulties to multiply plus or elite clones from them. The main cause for this phenomenon is attributed to polyphenols contained in those trees. In the present study, polyphenols contained in *E. camaldulensis* cuttings or dissolved in water after scission of the cutting shoots were identified and quantified to clarify the effects of polyphenols on rooting. In addition, assays of polyphenol oxidases from the cuttings were also tried to examine the involvement of the enzymes in rooting.

Results and Discussion

Figure 1 shows a total ion chromatogram of GC-MS analysis for the water-soluble extracts from cutting shoots of an old tree (Parent). Protocatechuic acid and gallic acid were identified as polyphenols in the water-soluble extracts and organic solvent extracts from both cutting shoots of young trees (Clone) and Parent. As shown in Table 1, however, detection patterns of the two polyphenols varied depending on the treatment and season when the cutting shoots were collected. When a polyphenol oxidase inhibitor, NaN_3, was used, both gallic acid and protocatechuic acid were detected in Clone, whereas only protocatechuic acid in Parent. In this system, the two polyphenols were quantified in the water-soluble extracts and organic solvent extracts (Table 2). Clone had more gallic acid than Parent, while Parent did more protocatechuic acid than Clone. However, this results were not enough to explain the effects of both polyphenols on the difference of rooting ability. Then polyphenol oxidase activity was assayed for Parent and Clone cutting shoots according to the method of Nozue *et al.* [1]. However, no activity were detected in either protein solutions probably because of inhibitional effects of polyphenols. In spite of no data on activity of polyphenol oxidases, the results obtained in the present study suggest that characteristic difference in the enzymes between Parent and Clone cutting shoots may affect the difference in rooting ability between them.

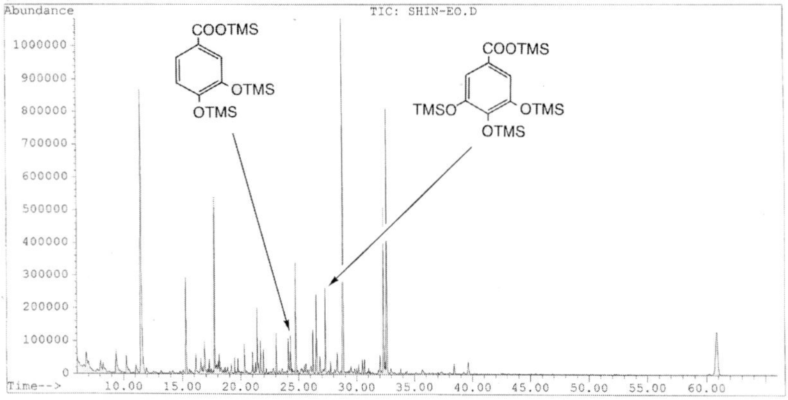

Figure 1 Total ion chromatogram of GC-MS analysis for the water-soluble extracts from Parent cutting shoots.

Table 1 Water-soluble polyphenols detected from cutting shoots by GC-MS analysis.

Treatment	Season	Plant	Gallic acid	Protocatechuic acid
Deep-frozen	Spring	Parent	d	d
Deep-frozen	Spring	Clone	d	nd
Without deep-frozen	Spring	Parent	nd	nd
Without deep-frozen	Spring	Clone	nd	nd
Without deep-frozen	Fall	Parent	nd	nd
Without deep-frozen	Fall	Clone	nd	nd
Without deep-frozen, NaN_3	Fall	Parent	nd	d
Without deep-frozen, NaN_3	Fall	Clone	d	d

d; detected. nd; not detected.

Table 2 Quantification of gallic and protocatechuic acids in the water-soluble and organic solvent extracts.

Extracts	Plant	Gallic acid (μmol / gFW)	Protocatechuic acid (μmol / gFW)
Water-soluble	Parent	–	0.26
	Clone	0.44	0.24
Organic solvent	Parent	2.91	1.53
	Clone	4.65	0.88

FW; fresh weight.

References

[1] Nozue, M., Souri, M., Arakawa, D. and Kojima, M. J. Plant. Physiol. 153 (1998) 552.

Stable Isotope Tracer Experiments Give Strong Evidence Supporting a Controversial Pathway in Lignin Biosynthesis

Kazuhiko Fukushima
Graduate School of Bioagricultural Sciences, Nagoya University, Nagoya 464-8601, Japan

Introduction

Lignin is composed mainly of three structural units: *p*-hydroxyphenyl (H), guaiacyl (G), and syringyl (S) moieties. The composition of lignin varies with the species and type of tissue, cell, and wall layer. Since the S/G ratio has a great effect on the quality of lignin, many attempts have been made to alter this ratio by genetic engineering. However, the manipulation of genes encoding enzymes of the conventional lignin pathway established by Higuchi [1] has generated unexpected results, which have led the scientific community to re-evaluate lignin biosynthesis. Therefore, our understanding of lignin biosynthesis has progressed rapidly in the last decade. Currently, the lignin biosynthetic pathway is believed to be a metabolic grid, and the focus of interest in lignin biosynthesis is how G intermediates are hydroxylated and methylated to the syringyl monolignol in angiosperms [2]. Our group investigated a controversial pathway in the grid using stable isotope-labeled tracer experiments to clarify whether the pathway functions. Here we explain the efficiency of the stable isotope tracer method and discuss recent findings using this method.

Theoretical basis for stable isotope tracer experiments combined with GC-MS

Investigating the pathway from coniferyl alcohol to sinapyl alcohol

The three monolignols differ only in the pattern of substitutions on the aromatic ring. A metabolic grid exists for modifying the aromatic ring and reducing their side chains in monolignol biosynthesis, and there are potentially a large number of routes (pathways) that might be involved in regulating the relative levels of these monolignols [3]. The terminal groups on the side chains of lignin precursors play important roles in determining these pathways. Studies of the behavior of the 9-hydrogen atoms in lignin precursors during the biosynthesis of monolignol should help us to understand some aspects of this process. As Fig. 1 shows, when living trees are fed labeled precursor (**1**), there are several possible pathways by which the precursor can be incorporated into lignin. It has been proposed that *O*-methylation of the aromatic ring can occur at the cinnamyl acid or cinnamoyl CoA ester stage. Therefore, the labeled precursor (**1**) could be incorporated into S lignin via the route **1** → **2** → **3** → **4** → **8** → **12** → **11** → **14** → **15** or via **1** → **2** → **3** → **7** → **11** → **14** → **15**. In these pathways, deuterium atoms at the 9-position of the side chains of the precursors are released when the alcohol is oxidized to an acid or a CoA ester. In such cases, trideuterium-labeled S units should be found in lignin as the DFRC monomer by GC-MS [4]. If the structural subunits of lignin can be converted at the monolignol stage via the route **1** → **5** → **9**, the pentadeuterium-labeled precursor should be incorporated into S lignin. Possible fates of the labeled lignin precursor (**1**) in newly formed xylem are shown in Fig. 1. Using the labeled precursor and tracing the distribution of the isotope among lignin subunits produced by the DFRC method, we can obtain information about the mechanisms of monolignol biosynthesis [5, 6].

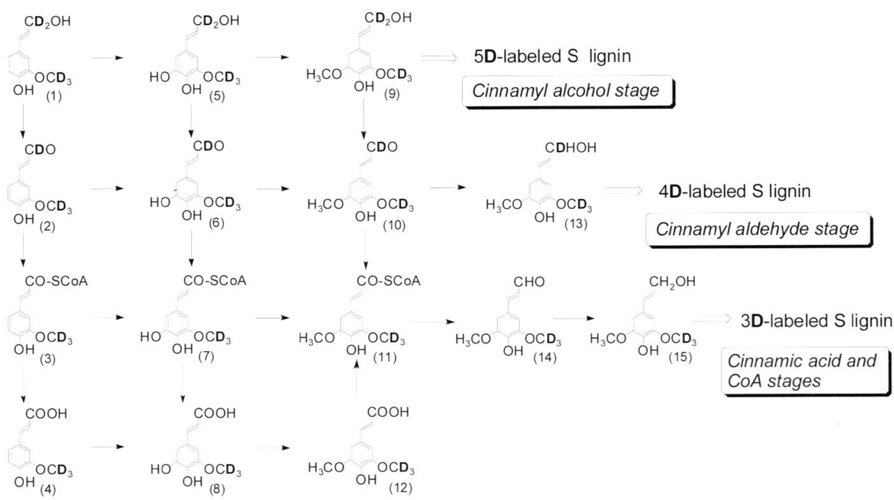

*Figure 1 Possible behavior of 5D-labeled coniferyl alcohol (**1**) during the formation of syringyl lignin.*

Investigating the pathway from sinapic acid to sinapyl alcohol

The sinapic acid-[8-^{13}C, 3,5-OCD_3] used in the tracer experiments was labeled with ^{13}C at the 8-position and with deuterium at both methoxy groups [7]. Methoxy deuterium in sinapic acid is very stable, and is not released during administration or subsequent chemical analysis [8]. As shown in Fig. 2, if syringyl moieties (DFRC monomer [4]) containing all labels (m+7) are detected by GC-MS, this suggests that labeled sinapic acid is converted to sinapyl alcohol via the proposed route, and it also supports the traditionally accepted pathway, in which sinapic acid is the precursor of syringyl lignin, as proposed by Higuchi [1]. If the syringyl moieties containing some of the label (m+4) are detected, it suggests that labeled sinapic acid has released one of the labeled methyl groups during lignin monomer biosynthesis. Conversely, if sinapic acid is not a precursor of syringyl lignin, no labeled syringyl lignin should be detected. The carbon at the 8 position was labeled with ^{13}C to detect the incorporated region rapidly by analyzing the isotope ratio (δ^{13}C) using mass spectrometry.

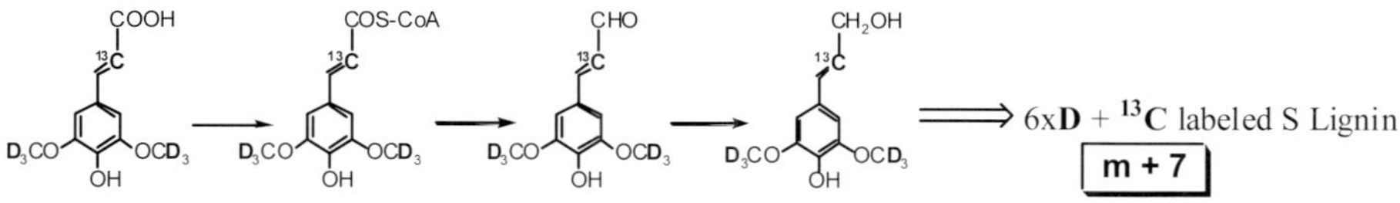

Figure 2 Possible pathway for the incorporation of sinapic acid-[$8^{-13}C$, 3,5-OCD_3] into syringyl lignin.

Results and Discussion

Conclusive evidence for the cinnamyl alcohol pathway

Radiolabeled coniferin was partially incorporated into S units in addition to G units in newly formed lignin in *Magnolia kobus* [9, 10]. This suggests the conversion from coniferyl alcohol to sinapyl alcohol. Feeding experiments with deuterium-labeled coniferyl alcohol (**1**) provided conclusive evidence for a novel alternative pathway for regulating the syringyl to guaiacyl ratio in lignin in *Magnolia* and *Eucalyptus* [5, 6]. This pathway (Fig. 1) from coniferyl alcohol (**1**) to sinapyl alcohol (**9**) might be composed of two steps [11], *i.e.*, hydroxylation and *O*-methylation, in addition to the cinnamate pathway. Since coniferin, which is both a storage and a transportable form of coniferyl alcohol, accumulates at an early stage in the differentiation of the xylem before lignification in some angiosperms, such as *Magnolia* [12], conversion *via* the cinnamyl alcohol pathway might play an important role in the synthesis of sinapyl alcohol.

In *Arabidopsis* mutants defective in ferulate 5-hydroxylase, the number of S units was below the detection limit [13], indicating that the pathways leading to the synthesis of sinapyl alcohol were blocked. However, this enzyme catalyzed the conversion of coniferyl aldehyde and coniferyl alcohol more effectively than did ferulate [14]. This 5-hydroxylase might also play an important role in the conversion of coniferyl alcohol to sinapyl alcohol.

Recently, it was found that *O*-methyltransferase (COMT I) from tobacco could *O*-methylate 5-hydroxyconiferyl alcohol [15], suggesting that COMT I, which catalyses the synthesis of S units *in planta*, operates at the level of the alcohol as well as the free acid and CoA ester.

The pathway via sinapic acid depends on the species

To clarify the biosynthetic pathway of sinapyl alcohol in angiosperms, tracer experiments with stable isotopes were performed using robinia (*Robinia pseudoacacia* L), oleander (*Nerium indicum* Mill.), magnolia (*Magnolia kobus* DC.), and Arabidopsis (*Arabidopsis thaliana*) species. The precursors used in the experiment were ^{13}C- and ^{2}H (D)-labeled ferulic acid-[8-^{13}C, 3-OCD_3] and sinapic acid-[8-^{13}C, 3,5-OCD_3]. The incorporation of labeled precursor into lignin (m+7) was confirmed by GC-MS analysis of the DFRC products [7]. Crude extracts of differentiating xylem or stem from these plants were also assayed for 4-coumarate-CoA ligase (4CL; EC 6.2.1.12) activity using sinapic acid and ferulic acid as substrates. In robinia and oleander, 4CL activity toward sinapic acid was detected, and labeled sinapic acids were incorporated into syringyl lignin. These results indicate that robinia and oleander have a pathway that produces sinapyl alcohol from sinapic acid via sinapoyl-CoA. By contrast, in magnolia and Arabidopsis, 4CL activity toward sinapic acid could not be detected, and labeled sinapic acid was not incorporated into lignin. These results suggest that syringyl lignin biosynthesis in angiosperms operates via multiple pathways that depend on the species [7].

Conclusion

The specially designed stable isotope labeled precursor described above enabled us to get useful information on the biosynthesis of lignin *in planta*, although lignin cannot be isolated as is and is a very complicated biopolymer lacking repeated linkage units.

References

[1] Higuchi, T. Wood Sci. Technol. 24 (1990) 23.
[2] Chapple, C. and Carpita, N. Current Opinion in Plant Biology 1 (1999) 179.
[3] Fukushima, K., J. Plant Research 114 (2001) 499.
[4] Lu, F. and Ralph, J., J. Agric. Food Chem. 45 (1997) 2590.
[5] Chen, F., Yasuda, S. and Fukushima, K. Planta 207 (1999) 597.
[6] Chen, F., Yasuda, S. and Fukushima, K. J. Wood Sci. 45 (1999) 487.
[7] Yamauchi, K., Yasuda, S., Hamada, K., Tsutsumi, Y. and Fukushima, K. Planta 216 (2003) 496.
[8] Yamauchi, K., Yasuda, S. and Fukushima, K. J. Agric. Food Chem. 50 (2002) 3222.
[9] Matsui, N., Fukushima, K., Yasuda, S. and Terashima. N. Holzforschung 48 (1994) 375.
[10] Matsui, N., Fukushima, K., Yasuda, S. and Terashima. N. Holzforschung 50 (1996) 408.
[11] Matsui, N., Chen, F., Yasuda, S. and Fukushima, K. Planta 210 (2000) 831.
[12] Fukushima, K., Taguchi, S., Matsui, N. and Yasuda, S. Mokuzai Gakkaishi 42 (1996) 1029.
[13] Chapple, C., Vogt, T., Ellis, B, E. and Somerville, C. R. Plant Cell 4 (1992) 1413.
[14] Humphreys, J. M., Hemm, M. R. and Chapple, C., Proc. Natl. Acad. Sci. U.S.A. 96 (1999) 10045.
[15] Maury, S., Geoffroy, P. and Legrand, M. Plant Physiol. 121 (1999) 215.

Characterization of Potassium Channels from *Arabidopsis thaliana*

Tatsuki Ibuki[a], Yukiko Tokida[a], Nobuyuki Matsuda[a], Katrin Czempinski[b], Bernd Muller-Roeber[b], Toshihiro Ona[c] and Nobuyuki Uozumi[a]

[a]*Bioscience and Biotechnology Center, Nagoya University, Nagoya 464-8601, Japan*
[b]*Universität Potsdam, Institut für Biochemie und Biologie, Karl-Liebknecht-Str. 24-25, Haus 20, D-14476 Golm, Germany and Cooperative Research Group of the Max-Planck-Institute of Molecular Plant Physiology, Am Mühlenberg 1, 14476 Golm, Germany*
[c]*Graduate School of Bioresource and Bioenvironmental Sciences, Kyushu University, Fukuoka 812-8581, Japan*

Introduction

Potassium ion (K^+) is the most abundant cation inside prokaryotic and eukaryotic cells and the K^+ transport system in biological membranes is essential for maintenance of membrane potential and diverse cellular processes in all organisms studied thus far. Bacterial K^+ uptake systems have been investigated extensively. In *E. coli*, three kinds of K^+ transporters have been identified as major K^+ transport uptake systems. *E. coli* Kdp is a P-type ATPase which has been classified as a high-affinity K^+ transport system. The Kup transporter in *E. coli* is a low affinity type of K^+ uptake transporter and the Trk system is the third K^+ transport system. In fact, an *E. coli* mutant, lacking the three K^+ uptake systems, Trk, Kup, and Kdp, could not grow on K^+-limiting medium. In Arabidopsis plants, the K^+ channels proteins exist [1]. More than ten genes encoding the homologous channels are dispersed in all of the genome and these homologous channels are expressed in Eucalyptus. We isolated the gene encoding the novel K^+ channel from Arabidopsis and characterized the function using *E. coli* expression system.

Experimental Procedures

Growth conditions - *E. coli* LB2003 which lacks the three K^+ uptake systems Trk, Kup, and Kdp was used as host strain for growth tests of the K^+ transport. The transformants of the *E. coli* mutants cells were cultured in the synthetic medium supplemented with required anitibiotics and indicated cations. The gene expression was induced by addition of 0.1 mM IPTG into the medium.

Isolation and expression of the K^+ channel genes –the gene encoding K^+ channels was amplified by PCR using *Arabidopsis thaliana* cDNA library as a template DNA. The PCR fragment was subcloned into the pPAB404.

Measurement of K^+ uptake -The K^+ influx measurement was essentially performed in Tolema et al. [2]. Cells were pre-cultured in synthetic medium (0.5 mM KH_2PO_4, 50 mM Tris-HCl (pH 7.0), 5 mM $(NH_4)_2SO_4$, 10 mM glucose, 6 μM $FeSO_4$, 0.4 mM $MgCl_2$) in the presence of 30 mM KCl, 0.5 mM IPTG, and 50 μg/ml Ampicillin. The pre-cultured cells were then collected by centrifugation and resuspended in synthetic medium with varying concentrations of KCl. Tris-EDTA treated cells were loaded with

NaCl. Net uptake of K^+ by these cells was measured by the silicone filtration technique. Contents of K^+ in the cell pellets were determined by atomic adsorption. For the correction of extra cytoplasmic cations in the cell pellet, the extra cytoplasmic space in this fraction was taken to amount to 1.25 μl/mg dry weight, and the concentrations of K^+ in this space were assumed to be equal to those in the medium. With this correction, net uptake of K^+ was approximately linear with time and extrapolated to about zero at t = 0. We therefore concluded that these assumptions are valid.

Results and Discussion

Two genes coding K^+ channel are isolated from Arabidopsis cDNA library and are put under the control of t*ac* promoter in the vector which can be replicated in *E. coli.* The deduced protein included a glycine-tyrosine-glycine (GYG) motif, which is essential to express K^+ transport and ion selectivity. The *E. coli* LB2003, lacking the K^+ uptake systems, Trk, Kup and Kdp, could not grow on K^+ limiting medium. The growth of cells with empty vectors in medium containing 15 mM K^+(Figure 1). In contrast, the cells expressing one of the higher plant K^+ channels genes grew up well, but the cells expressing the other gene did not grow on the same medium. These results suggested that the K^+ channel was both translocated into the cell membrane and expressed with inward transporting K^+ activity. The K^+ uptake transport was measured by atomic adsorption (Fig. 2). We confirmed the K^+ influx in the cells expressing the channel but the transport activity was relative low. The C-terminal region of the channel which shows the complementation for the mutations of *E. coli* K^+ uptake activity contains the putative calcium binding domain. When the C-terminal containing the domain was removed, the channel still complement the mutation of the *E. coli*, indicating the domain may not involve regulation on the channel activity.

K^+ channel

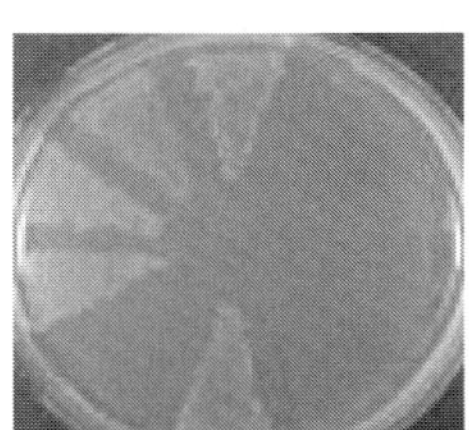

Non-functional channel

Chimera

Vector

Figure 1 Complementation of the K^+ uptake activity mutations in E. coli *by isolated gene encoding K^+ channels.*

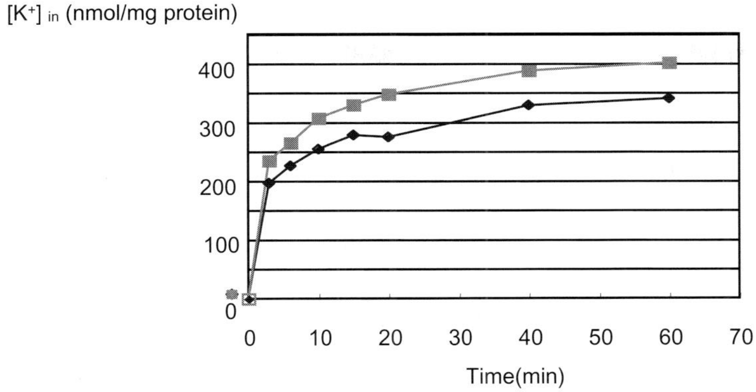

Figure 2 K^+ uptake activity of the cells expressing the K^+ channels.

We constructed the chimera proteins between the functional channel and non-functional channel. The chimera protein which has functional channel at the N terminal half and non-functional channel at the C-terminal half conferred the K^+ uptake activity in the *E. coli* expression assay (Figure 1). We are now in progress to narrow down the region which is responsible for the K^+ channel uptake activity.

To determine the localization of the channel in plant, the promoter-GUS reporter gene was introduced into the Arabidopsis and the antibody has been generated. We will discuss the role of the channel in plant.

References

[1] Uozumi, N., Am. J. Physiol. 281 (2001) C733.
[2] Tolema, N., Bakker EP., Suzuki A. and Nakamura, T., FEBS Let. 450 (1999) 217.
[3] Uozumi, N., Yamada, K., Goshima, S., Ona, T. and Oiki, S. J. Membrane Biol. 181 (2001) 163.
[4] Sato, Y., Sakaguchi, M., Goshima, S., Nakamura, T. and Uozumi, N. J. Biol. Chem. 278 (2003) 13227.

A New Combination Device Comprised of Surface Plasmon Resonance and Fluorescence Microscopy for a Rapid Screening of Anticancer Phenolic Compounds

Atsushi Kosaihira[a], Takashi Fukumori[b], Kokki Sakai[a] and Toshihiro Ona[a]

[a]Graduate School of Bioresource and Bioenvironmental Sciences, Kyushu University, Hakozaki, Higashi-ku, Fukuoka 812-8581, Japan

[b]Faculty of Agriculture, Kyushu University, Hakozaki, Higashi-ku, Fukuoka 812-8581, Japan

Introduction

Recently, physiological activity of phenolic compounds received significant attention; for example; antiproliferative effect on cancer cells has been reported in quercetin [1], which is contained in a bark of *Eucalyptus* species and in a propolis [2, 3]. However, the mechanism of phenolic compounds against cells is very complicate. Furthermore, screening of phenolic compounds is time-consuming because it is generally performed by the proliferarive inhibition of given cells in vitro. A new rapid method that enables to understand both mechanism of action and practical effect of phenolic compounds in a system is expected.

Various kinds of vibrational spectroscopic methods contribute to the non-destructive analysis of biological phenomenon. Recently, Fourier transform infrared spectroscopy and near infrared Raman spectroscopy are applied to find cancer tissues [4, 5]. On the contrast, little has been explored for rapid screen of anticancer drugs using living cells partly because of the fluorescence disturbance emitted from cells and also the damage of cells caused by light irradiation. On the other hand, the surface plasmon resonance sensor (SPR) enables to detect interactions between living cells and supplied molecules without the problems [6]. However, detected signals have to be explained by other offline methods such as fluorescence microscopy.

In this study, we report the development of a new combination device for a rapid screening of anticancer drugs using living cells.

Results and Discussion

The device consists of an SPR (under the cells) and a fluorescence microscope (over the cells). The cells cultured on a sensor chip coated with gold in 50 nm thickness are simultaneously measured by SPR and a fluorescence microscope. We examined the reaction of cancer cells, highly malignant pancreatic cancer line, MIA PaCa-2 to quercetin (Figure 1) as an example of anticancer drug [4].

As a result, the exposure of quercetin to cancer cells induced increase in SPR signal (Figure 2). The maximal increase in SPR signal induced by 100 μM quercetin was approximately 0.2°.

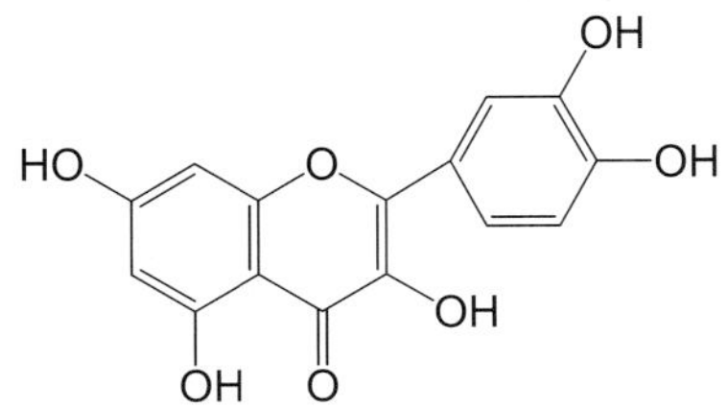

Figure 1 Chemical structure of quercetrin.

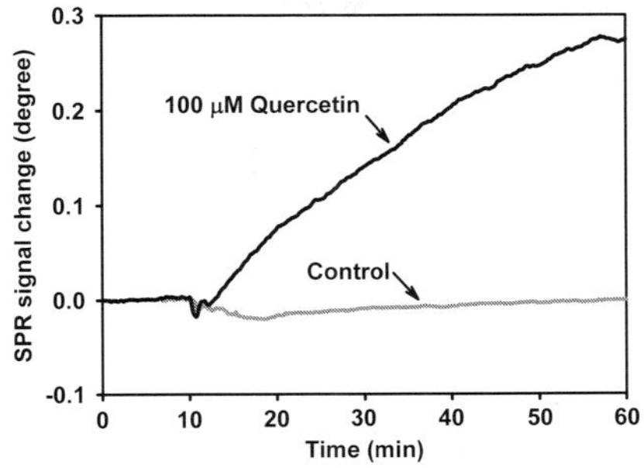

Figure 2 SPR signal changes detected from MIA PaCa-2 cells.

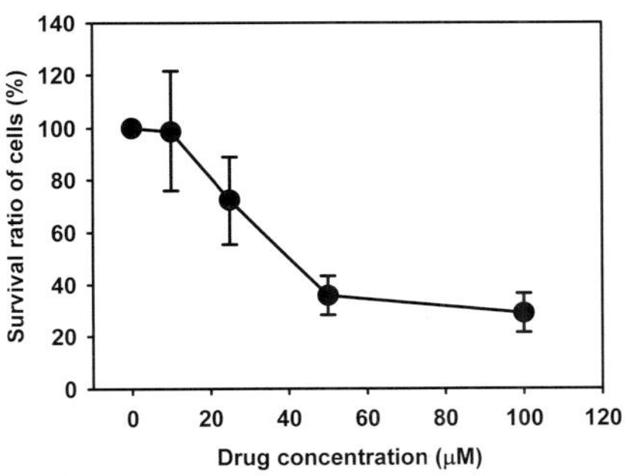

Figure 3 Antiproliferative effect of quercetin on MIA PaCa-2 cells.

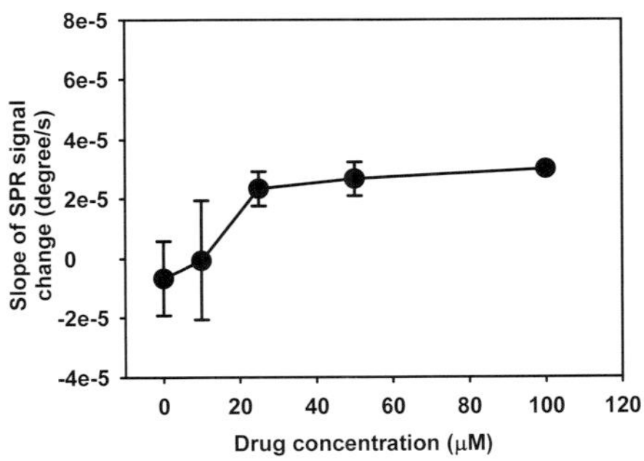

Figure 4 Effect of quercetin concentration on a slope of SPR signal.

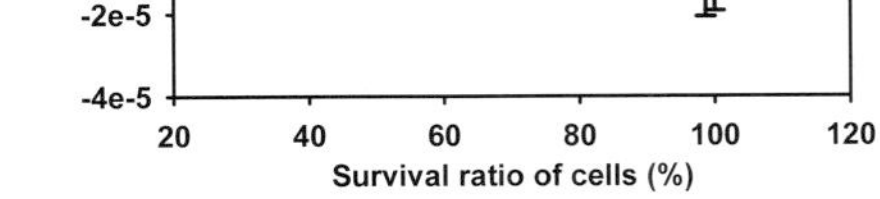

Figure 5 Relationship between Antiproliferative effect and a slope of SPR signal of cancer cells exposed to quercetin.

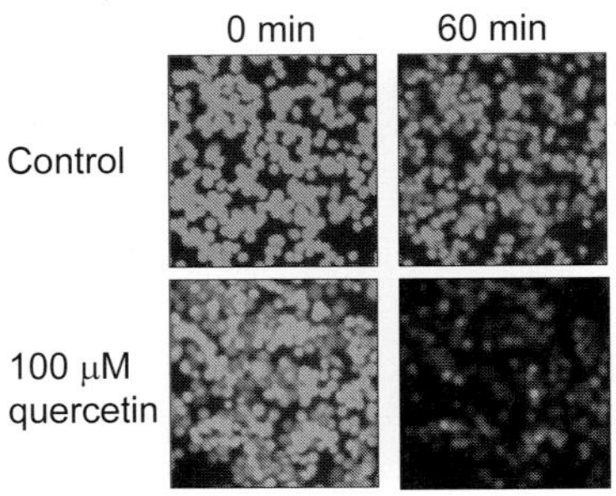

Figure 6 Fluorescent images of cancer cells before and after exposed to quercetin.

Furthermore, there was significantly high relationship between the values of linear increase in SPR signal sampled for 300 sec and the conventionally measured survival ratio of cancer cells after 48 hours, both exposed to various concentration of quercetin (Figures 3-5). The measurement of fluorescent intensity with 5-(and-6)-carboxy SNAFL-1 diacetate was simultaneously performed by a fluorescence microscope for quercetin and revealed decrease in cell activity relating to the SPR signal change (Figure 6). Consequently, this newly developed device will contribute to screen new anticancer drug, rapidly.

Reference

[1] Mouria, M., Gukovskaya, A. S., Jung, Y., Buechler, P., Hines, O. J., Reber, H. A. and Pandol, S. J. Int. J. Cancer 98 (2002) 761.

[2] Conde, E., Cadahía, E., Díez-Barra, R. and García-Vallejo, M. C. Holz Roh. Werk. 54 (1996) 175.

[3] Suzuki, I., Hayashi, I., Takai, T., Groveman, D. S. and Fujimiya, Y. Cancer Biother. Radio 17 (2002) 553.

[4] Erukhimovitch, V., Talyshinsky, M., Souprun, Y. and Huleihel, M. Photochem. Photobiol. 76 (2002) 446.

[5] Kaminka, S., Ito, T., Yamazaki, H., Kohda, E. and Hamaguchi, H. J. Raman Spectrosc. 33 (2002) 498.

[6] Hide, M., Tsutsui, T., Sato, H., Nishimura, T., Morimoto, K., Yamamoto S. and Yoshizato, K. Anal. Biochem. 302 (2002) 28.

The Feasibility of a Biotest Battery Applicability to Construct an Early Warning System for Leachate from Solid Wastes and Recycled Materials

Shino Mohri [a], Ryo Shoji [b] and Masato Yamada [a]
[a]Research Center for Material Cycles and Waste Management, National Institute for Environmental Studies, 16-2 Onogawa, Tsukuba, Ibaraki 305-8506, Japan
[b]Department of Chemical Science and Engineering, Tokyo National College of Technology, 1220-2 Kunugida-Machi, Hachioji, Tokyo 193-0997, Japan

Introduction -an early warning system-

In Japan, public concerns become more sensitive against the construction and operation of landfill sites since the leachates from them were reported as hazardous chemicals for human and ecosystem by the mass media. The chemical analysis can identify the presence of toxic compounds but not all in leachate to assess the risk because of the limitation of applied analytical methods. Furthermore, the analytical cost is a serious burden for landfill owners and managers.

The toxicity tests using bioassay, i.e. bioassays or biotests directly measuring toxic effects on organisms, can reflect the cumulative and synergistic effects of all the compounds including unknown species in leachate. The objective of this research project is to show the feasibility of cost-effective biotest batteries to monitor landfill leachate (Figures 1 and 2).

We detected various toxic potencies, namely acute toxicity, genotoxicity, biomarkers and cytotoxicity in untreated and treated leachates from municipal and industrial solid waste disposal sites. Using the same samples, chemical analysis was performed for conventional water quality parameters, heavy metals and some organic compounds. The combined biological and chemical analysis data have provided valuable information of potential impacts of the leachate on the surrounding water environment and of the directions for the landfill management based on the precautionary principle. To assess the applicability of various biotests in the practical management of landfill sites, cost, simplicity and sensitivity of each test are also compared with each other. Some chemical data showed the certain relationships between chemical characteristics of leachate and the toxicities, but not all, which implies the difficulties in estimating toxic hazards from leachate by chemical analysis only.

In addition, to construct a reasonable biotest battery, we attempted to hierarchical scoring of the test results obtained (Tables 1 and 2). The present study demonstrated the feasibility of a biotest battery examined for the practical use under various conditions of final target, cost, location and so on. The potential of batteries composed of cost-effective simple biotests was also examined as attractive alternatives to expensive conventional bioassays for routine monitor of leachates.

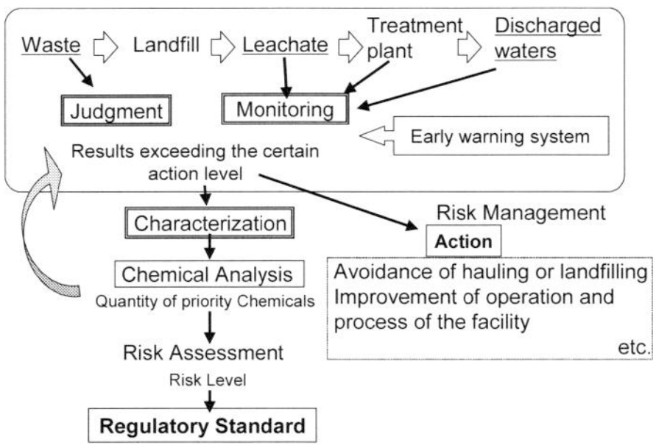

Figure 1 Early warning system for risk management of leachates from landfill sites.

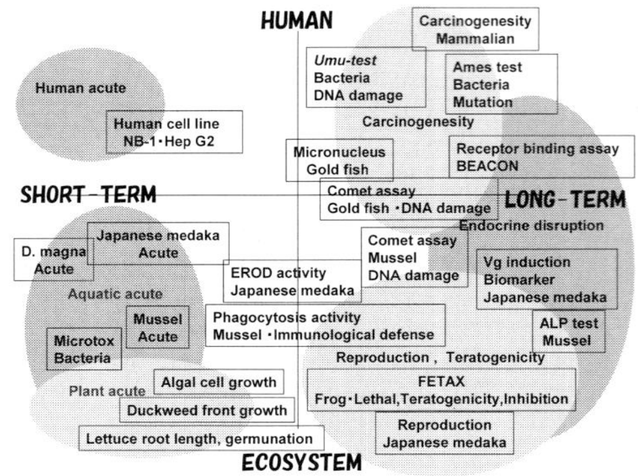

Figure 2 Mapping of toxicity tests for endpoints and objectives.

Table 1 Scoring designs of test results for a biotest battery. (EC50; Median effective concentration, LC50; Median lethal concentration)

Genotoxicity , Biomarker	Score
0.05< p	1
0.01< p<0.05	2
p<0.01	3

Cytotoxicity , Ecotoxicity	Score
EC50,LC50 > 100%	1
100%> EC50,LC50 >10%	2
10% > EC50,LC50	3

Table 2 Results of leachate samples from landfill sites (A-G; landfill site name, -1;Rraw leachate, -2; Treated reachate from the water treatment process).

Endpoint	Bioassay / Sample name	A-1	A-2	C-1	C-2	D-1	D-2	D-3	D-4	E-1	E-2	F-1	F-2	G-1	G-2	Notes and reference	
Cytotoxicity	Hep G2: survival	1	1	1	1	2	1	1	2	1	1	1	1	1	1	Human hepatocellular carcinoma cell lines	[1]
	NB-1: survival	1	1	1	1	2	2	2	2	1	1	2	1	2	2	Human neuroblastoma cell lines	[2]
Genotoxicity	Bacteria:*umu*-test(-S9)	2	1	3	2	2	2	1	2	2	1	2	1	2	1	*Salmonella typhimurium* TA1535/pSK1002	[3]
	Bacteria:*umu*-test(+S9)	3	1	3	2	2	2	3	2	2	1	3	2	2	2		
	Goldfish: Comet	2	2	2	2	3	2	3	3	3	2	3	3	2	2	Alkaline single cell gel (comet) assay	[4]
	Mytilus: Comet	2	1	1	1	3	2	-	-	3	2	1	1	3	3		
	Goldfish: MN: blood	3	2	2	1	1	1	1	1	1	1	1	1	1	1	Micronucleus test	[5]
	Goldfish: MN: gill	2	2	2	2	3	2	3	3	2	2	3	2	2	2		
Biomaker	Mytilus: phagocytosis	2	2	1	2	3	2	-	-	1	2	3	2	1	3		[6]
	Medaka: EROD	2	2	2	2	2	3	3	2	2	2	3	1	2	3	Cytochrome P450 1A (CYP1A) induction	[7]
	Medaka: Vtg	2	2	2	3	2	1	2	2	1	1	1	2	2	2	Vtelloghenin(Vtg)induction	[8]
Acute lethality etc.	Mytilus: lethality	1	1	2	1	3	1	-	-	1	1	2	1	1	1		/
	FETAX	4	1	5	2	5	5	5	5	2	3	5	2	4	4	Frog Embryo Teratogenesis Assay *Xenopus*	[9]
	Medaka: lethality	3	1	3	1	3	2	3	3	1	1	3	1	1	1	OECD TG203	/
	Thamno: lethality *	3	1.1	20	1.9	2.2	0	0	3.7	1	0	14	0	4	3.3	Thamnotoxkit F(*Thamnocepharus platyurus*)	[10]
Algae glowth inhibition	Algae: cell growth *	3.4	1.1	17	1.3	35	3.7	24	46	0	0	9.1	1.4	0	0	*Selenastrum capricornutum*(NIES35)	/
	Algae: cell growth	-	-	-	-	-	-	-	-	-	-	-	-	3	1	OECD TG201	
	Algaltox: cell growth	-	-	-	-	-	-	-	-	3	2	3	1	3	3	TOXKIT Algal beads	/
Plant	Duckweed: frond growth *	5.6	2.4	12	0	26	6.3	4.3	29	1.4	0	9.1	2.8	3.8	4	*Lemna minor* 1769 OECD TG221	[11]
	Lettuce: root length *	0	0	2	1.2	8.3	6.7	2.4	10	0	3.1	2.5	3.4	1.1	5.3	*Lactuca sativa*	[12]
	Lettuce: germination *	1.3	1.1	1.1	0	2	1.7	0	2	0	0	1.2	1.4	0	1.2	*Lactuca sativa*	[13]
Inhibition	ToxAlert: luminescence	-	-	-	-	-	-	-	-	1	1	2	1	1	1	ToxAlert 100 system(*Vibrio fischer*)	/

- ; Not measured, * ; Toxicity Unit (TU) = 1/EC50(%) x 100.

Application to solid wastes

At the landfill sites of solid wastes in Japan, leaching test , which is Japanese leaching teat 13 (JLT13),and chemical analysis, which aimed to decrease toxicity mainly of heavy metals, are conducted to judge if carried wastes can be disposed. From a view of whole toxic evaluation, no bioassay was standardized for the waste management in Japan, and no system is available to evaluate the required treatment process to reduce toxic substances, which was demonstrated by the TIE (Toxicity Identification Evaluation) /TRE (Toxicity Reduction Evaluation) in U. S. EPA [14-16]. On the other hand, our research center suggests the ways in performing appropriate material cycles and waste management to construct a material cycle oriented society as a system including the policies of "waste prevention" and "recycling". From them, we started to refine this early warning system and to modify the leaching method suitable to evaluate recycled materials for the provision of more safety and wide (accelerated) use of recovered materials.

We examined elute-ability and toxicity of organic compounds in several industrial wastes and recovered materials, such as fry ash, bottom ash, shredder dust, construction waste, gypsum board, wood waste and dewatered sludge by both standard and modified leaching tests. The eluate by leaching test was analyzed for toxicity, total organic carbon (TOC) and heavy metals. From the results of several leaching tests with various conditions, TOC levels were greatly varied by the conditions, and aqueous leaching tests were suitable for both chemical and biological testing. The toxicity was assessed by *D. magna* (OECD TG202), algae (OECD TG 201) and human cell line Hep G-2. The some eluates by JLT13 showed toxicity as in Figure 3. And some diverse effects such as over growth in algal growth inhibition test, expected the modification of the leaching test system. Toxicities observed in eluates weren't derived from salinity and heavy metals. Further bioassays revealed some toxicities, which were explained by the presence of organic compounds.

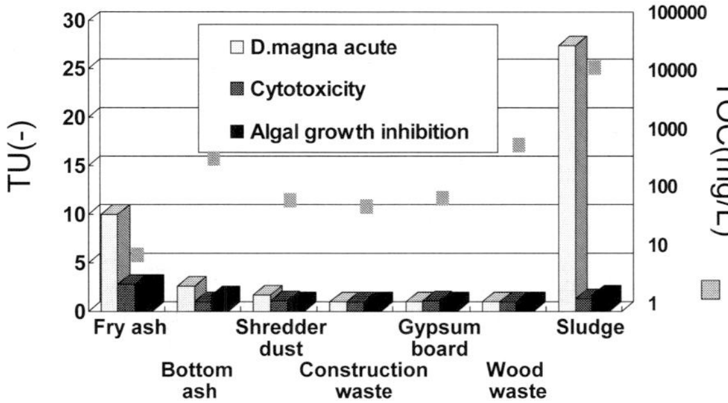

Figure 3 The results of bioassays of eluates by leaching test with solid samples. Toxicity Unit (TU) = 1/EC50(%) x 100.

Conclusions

Toxicity tests have a long history in drug and chemical industries to ensure safety of their products guided by the regulation law, and wide variety of methods are available for various toxicity endpoints, namely acute, chronic, carcinogenic, mutagenic and genotoxic matters. In environmental field in Japan, these tests have been applied only to prosecute undecided and sporadic risks in environment and emissions. In this report, we intend to add more positive utility values on toxicity tests for the protection of our environment. More detailed methodology and technology, which need to realize the early warning system, is now developing in our department corroborating with some universities and institutions. We call for more interdisciplinary discussions from all related fields.

Acknowledgement

The authors thank the experimental collaboration for A. Yasuhara (chemical analysis), Y. Inoue (priority list) and H. Takigami (FETAX) in NIES (Japan), A. Sakoda and Y. Sakai (TIE/TRE) in Institute of Industrial Science, University of Tokyo, Japan, N. Kinae (goldfish) in University of Shizuoka, Japan, Y. Ono (Medaka fish) in University of Okayama, Japan, T. Kusui (Mussel), Toyama Prefectural University, Japan, M. Kunimoto (micro biotest), Kitasato University, Japan, H. Okamura (ecotoxicity), Kobe University of Mercantive Marin, Japan and also for managers and researchers of waste management division at sampling sites.

Reference

[1] Scheers EM, Ekwall B, Dierickx PJ., *In vitro* long-term cytotoxicity testing of 27 MEIC chemicals on Hep G2 cells and comparison with acute human toxicity data, Toxicol *In Vitro*. 15(2) (2001)153-61.

[2] Ekwall, B., Kunimoto, M., MEIC evaluation of acute systemic toxicity: Part VI. The prediction of human toxicity by rodent LD50 values and results from 61 *in vitro* methods, ATLA, 26(1998)617-658.

[3] Oda,Y.,S.Nakamura,I., Evaluation of the new system(*umu*-test) for the detection of environmental mutagens and carcinogens,Mutation Research, 147(1985)219-229.

[4] R.Pandrangi *et al*., Alkaline single cell gel (comet) assay and genotoxicity monitoring using bullheads and carp assay, Environ.Mol.Mutagen.26(4) (1995)345-356.

[5] T.Tanisho *et a*l.: Monitoring of coastal water contaminated with mutagens and/or carcinogens using micronucleus test in fish, Environ. Mutagen Res.20(1998)1-9.

[6] Wan, C. P., Park C.S., and Lau B.H.S., A rapid and simple microfluorometric phagocytosis assay,Journal of Immunologocal Methods. 162(1993)1-7

[7] J. J. Whyte, R. E. Jung C. J. Schmitt, Ethoxyresorufin-*O*-deethylase(EROD) activity in fish as a biomarker of chemical exposure, Critical Rev. in Toxicol. 30(4)(2000)347-570.

[8] J. P. Sumpter, S. Jobling, Vitellogenesis as a biomarker for estrogenic contamination of the aquatic environment, Environmental Health Perspectives 103(1995)173-178.

[9] Standard guide for conducting the frog embryo teratogenesis assay- *Xenopus* (FETAX). Annual Book of ASTM Standards (1993)E-1439-91.

[10] Persoone,G., Development and validation of Toxkit microbiotests with invertebrates, in particular crustaceans. Chapter 30, Microscale Testing in Aquatic Toxicology. Eds by Wells,P.G., Lee,K, and Blaise,C., CRC Press(1998) 437-449.

[11] OECD *Lemna* sp., Growth inhibition test. OECD TG 221 (2000)

[12] Wong, W., Root elongation method for toxicity testing of organic and inorganic pollutants. Environ. Toxicol. Chem. 6(1987) 409-414.

[13] Wong, W., Comparative seed germination tests using ten plant species for toxicity assessment of a metal engraving effluent sample. Water,Air,Soil Pollut. 52(1990) 369-376.

[14] U. S. EPA. "Methods for Aquatic Toxicity Identification Evaluations, Phase I Toxicity Characterization Procedures, 2nd ed." (1991) EPA/600/6-91/003.

[15] U. S. EPA. "Methods for Aquatic Toxicity Identification Evaluations, Phase II Toxicity Identification Procedures for Samples Exhibiting Acute and Chronic Toxicity" (1993) EPA/600/R-92/080.

[16] U. S. EPA. "Methods for Aquatic Toxicity Identification Evaluations, Phase III Toxicity Confirmation Procedures for Samples Exhibiting Acute and Chronic Toxicity", (1993) EPA/600/R-92/003.

List of Authors

Da Silva Perez, Denilson

AFOCEL
Wood-Process Laboratory
Domaine de l'Etançon, AFOCEL, Nangis 77370
France
+33-1-60-67-02-50
+33-1-60-67-02-56
denilson.dasilvaperez@afocel.fr

Fukumori, Takashi

Kyushu University
Department of Forest and Forest Products Sciences
Faculty of Agriculture
6-10-1 Hakozaki, Higashi-ku, Fukuoka 812-8581
Japan
+81-92-642-2990
+81-92-642-2990
taka618@agr.kyushu-u.ac.jp

Fukusima, Kazuhiko

Nagoya University
Department of Applied Biological Sciences
Graduate School of Bioagricultural Sciences
Furo-cho, Chikusa-ku, Nagoya, Aichi 464-8601
Japan
+81-52-789-4160
+81-52-789-4163
kazu@nuagr1.agr.nagoya-u.ac.jp

Gellerstedt, Göran

Royal Institute of Technology, KTH
Department of Fibre and Polymer Technology
Drottning Kristinas Vag 53, Stockholm S-100 44
Sweden
+46-8-790-8109
+46-8-790-6166
ggell@pmt.kth.se

Haisaki, Hironori

Kyushu University
Department of Forest and Forest Products Sciences
Faculty of Agriculture
6-10-1 Hakozaki, Higashi-ku, Fukuoka 812-8581
Japan
+81-92-642-2990
+81-92-642-2990
hhaisaki@agr.kyushu-u.ac.jp

Ishida, Yasuyuki

Nagoya University
Research Center for Advanced Energy Conversion
Furo-cho, Chikusa-ku, Nagoya, Aichi 464-8603
Japan
+81-52-789-4666
+81-52-789-4666
yuki@apchem.nagoya-u.ac.jp

Ishiguri, Futoshi

Utsunomiya University
Department of Forest Science
Faculty of Agriculture
350 Mine-machi, Utsunomiya, Tochigi 321-8505
Japan
+81-28-649-5541
+81-28-649-5545
ao2m3701@cc.utsunomita-u.ac.jp

Kojima, Yasuo

Hokkaido University
Department of Forest Science
Graduate School of Agriculture
N9 W9, Kita-ku, Sapporo, Hokkaido 060-8589
Japan
+81-11-706-2526
+81-11-706-4180
koji@for.agr.hokudai.ac.jp

Kosaihira, Atsushi

Kyushu University
Department of Forest and Forest Products Sciences
Graduate School of Bioresource and Bioenvironmental Sciences
6-10-1 Hakozaki, Higashi-ku, Fukuoka 812-8581
Japan
+81-92-642-2990
+81-92-642-2990
atsu-kos@agr.kyushu-u.ac.jp

Mikame, Keigo

Mie University
Faculty of Bioresources
1515 Kamihama-cho, Tsu, Mie 514-8507
Japan
+81-59-232-1211 ext. 2541
+81-59-231-9517
mikame@bio.mie-u.ac.jp

Mizumoto, Miho

Kyushu University
Department of Forest and Forest Products Sciences
Graduate School of Bioresource and Bioenvironmental Sciences
6-10-1 Hakozaki, Higashi-ku, Fukuoka 812-8581
Japan
+81-92-642-2990
+81-92-642-2990
mihorin@agr.kyushu-u.ac.jp

Mohri, Shino

Research Center for Material Cycles and Waste Management
National Institute for Environmental Studies
Onogawa16-2, Tsukuba, Ibaraki 305-8506
Japan
+81-29-850-2805
+81-29-850-2016
mohri.shino@nies.go.jp

Murakami, Shuichi

Kyushu University
Department of Forest and Forest Products Sciences
Faculty of Agriculture
6-10-1 Hakozaki, Higashi-ku, Fukuoka 812-8581
Japan
+81-92-642-2990
+81-92-642-2990
shumura@agr.kyushu-u.ac.jp

Nazhad, Mousa M

School of Environment, Resources and Development
Asian Institute of Technology
P.O.Box 4, Klong Luang
Pathumuthani 12120
Thailand
+66-2-524-6687
+66-2-524-6681
m2na2i2@ait.ac.th

Nozaki, Hideo

Kyushu University
Department of Forest and Forest Products Sciences
Faculty of Agriculture
6-10-1 Hakozaki, Higashi-ku, Fukuoka 812-8581
Japan
+81-92-642-2990
+81-92-642-2990
pseudo@agr.kyushu-u.ac.jp

Obata, Yoshihiro

National Institute of Advanced Industrial Science and Technology
Advanced Wood-based Material Technology Group
Institute for Structural and Engineering Materials
2266-98 Anagahora, Shimoshidami, Moriyama-ku, Nagoya 468-8560
Japan
+81-52-736-7323
+81-52-736-7400
y-obata@aist.go.jp

Ohara, Seiji

Forestry and Forest Products Research Institute
Department of Forest Chemistry
1 Matsunosato, Tsukuba, Ibaraki 305-8687
Japan
+81-29-873-3211ext.525 or 453
+81-29-873-3797
oharas@ffpri.affrc.go.jp

Ohshima, Jyunichi

Utsunomiya University
Department of Forest Science
Faculty of Agriculture
350 Mine-machi, Utsunomiya, Tochigi 321-8505
Japan
+81-28-649-5539
+81-28-649-5545
jyun.o@d9.dion.ne.jp

Okayama, Takayuki

Tokyo University of Agriculture and Technology
Graduate School of Agriculture
Department of Environmental and Natural Resource Science
3-5-8 Saiwai-cho, Fuchu, Tokyo 183-8509
Japan
+81-42-367-5725
+81-42-334-5700
okayama@cc.tuat.ac.jp

Okuyama, Takashi

Nagoya University
Department of Applied Biological Sciences
Graduate school of Bioagricultural Sciences
Furo-cho, Chikusa-ku, Nagoya, Aichi 464-8603
Japan
+81-52-789-4151
+89-52-789-4150
tok@agr.nagoya-u.ac.jp

Ona, Toshihiro

Kyushu University
Department of Forest and Forest Products Sciences
Graduate School of Bioresource and Bioenvironmental Sciences
6-10-1 Hakozaki, Higashi-ku, Fukuoka 812-8581
Japan
+81-92-642-2989
+81-92-642-2989
ona@agr.kyushu-u.ac.jp

Sakai, Kokki

Kyushu University
Department of Forest and Forest Products Sciences
Graduate School of Bioresource and Bioenvironmental Sciences
6-10-1 Hakozaki, Higashi-ku, Fukuoka 812-8581
Japan
+81-92-642-2988
+81-92-642-2988
ksakai@agr.kyushu-u.ac.jp

Schimleck, Laurence

University of Georgia
Warnell School of Forest Resources
Athens, GA 30605
U.S.A.
+1-706-583-0464
+1-706-542-8356
lschimleck@smokey.forestry.uga.edu

Seino, Teruyuki

NEDO Fellow, Institute for Environmental Management Technology
AIST Tsukuba West, 16-1 Onogawa, Tsukuba, Ibaraki 305-8569
Japan
+81-29-861-8785
+81-29-861-8308
teruyuki-seino@aist.go.jp

Tateishi, Mari

Kyushu University
Department of Forest and Forest Products Sciences
Graduate School of Bioresource and Bioenvironmental Sciences
6-10-1 Hakozaki, Higashi-ku, Fukuoka 812-8581
Japan
+81-92-642-2990
+81-92-642-2990
maruko@agr.kyushu-u.ac.jp

Uozumi, Nobuyuki

Nagoya University
Bioscience and Biotechnology Center
Furo-cho, Chikusa-ku, Nagoya, Aichi 464-8601
Japan
+81-52-789-5202
+81-52-789-5206
uozumi@agr.nagoya-u.ac.jp

Wang Lili

Zhejiang University of Technology
College of Chemical Engineering and Material
Zhaohui 6 qu, Hangzhou, 310014, PRC
China
+86-571-88320416
lili_wang89@hotmail.com

Yamamoto, Hiroyuki

Nagoya University
Department of Applied Biological Sciences
Graduate school of Bioagricultural Sciences
Chikusa, Nagoya 464-8601
Japan
+81-52-789-4152
+81-52-789-4150
hiro@agr.nagoya-u.ac.jp

Yokota, Shinso

Utsunomiya University
Department of Forest Science
Faculty of Agriculture
350 Mine-machi, Utsunomiya, Tochigi 321-8505
Japan
+81-28-649-5539
+81-28-649-5545
yokotas@cc.utsunomita-u.ac.jp

Yoshizawa, Nobuo

Utsunomiya University
Department of Forest Science
Faculty of Agriculture
350 Mine-machi, Utsunomiya, Tochigi 321-8505
Japan
+81-28-649-5541
+81-28-649-5545
nobuoy@cc.utsunomita-u.ac.jp

Subject Index

S

T